W9-CAE-493

WIVES OF STEEL

Karen Olson

WIVES OF STEEL

Voices of Women from the Sparrows Point Steelmaking Communities

The Pennsylvania State University Press | University Park, Pennsylvania

LIBRARY OF CONGRESS
CATALOGING-IN-PUBLICATION DATA

Olson, Karen, 1943–
Wives of steel : voices of women from the Sparrows Point steel-making communities / Karen Olson.
p. cm.
Includes bibliographical references and index.
ISBN 0-271-02685-5 (alk. paper)
1. Wives—Effect of husband's employment on—Maryland—Sparrows Point—History.
2. Iron and steel workers—Maryland—Sparrows Point—History.
3. Steel industry and trade—Maryland—Sparrows Point—History.
4. Work and family—Maryland—Sparrows Point—History.
I. Title.

HQ1439.S63O46 2005
306.872'3'0975271—dc22
2005021392

Printed in the United States of America
Published by
The Pennsylvania State University Press,
University Park, PA 16802-1003

The Pennsylvania State University Press
is a member of the Association of
American University Presses.

It is the policy of
The Pennsylvania State University Press
to use acid-free paper. This book is printed
on Natures Natural, containing 50% post-consumer waste, and meets the minimum
requirements of American National
Standard for Information Sciences—
Permanence of Paper for Printed Library
Material, ANSI Z39.48–1992.

Frontispiece: Turner Station and Dundalk
activists, 2005. Photo by Marge Neal.

CONTENTS

ILLUSTRATIONS

ACKNOWLEDGMENTS

I owe my biggest debt to the women and men in Turner Station and Dundalk who shared personal stories, insights, and rich knowledge of the history of their community. From them I learned that making steel requires exceptional skill and an extraordinarily strong work ethic, and that none of the stereotypes we have about what steelworkers and the wives of steelworkers are like has any resemblance to the intellectual and social complexity of the Sparrows Point steel mill and community over the last 115 years. I also learned that building an enduring community requires commitments and sacrifices on behalf of goals that extend beyond individual well being. Indeed, the work of sustaining a community is never done.

Between 1985 and 2005, I interviewed more than eighty-five people about the Bethlehem Steel plant at Sparrows Point, and about the communities of Sparrows Point, Turner Station, and Dundalk. These ethnographic interviews were confidential, therefore only a few real names appear in this account. There are some individuals who are published writers or public figures; their real names are included.

Interviews with the following people made this book possible: Dee Anderson, Mort Anderson, Randy Anderson, Tyrone Beasley, Linda Beavers, Ann Bonner, Carol Brooks, Francis Brown, Kathleen Brown, George Burke III, Walter Burrell, Edie Butler, Herb Clark, Terry Coburn, Steve Craddock, Carla Crisp, Lois Hilton Crossett, Debbie Dillon, Charles Durant-Bey, Chuck Edgehill, Jerry Ernest, Janice Evans, Juanita Everly, Margaret Farrow, Frank Fisher, Willa Fisher, Paula Fleming, John Fournelle, Sally Frank, Scotty Fraser, Kathy Garrison, Frank Germani, Landon Godsey, Kathleen Graves, Ahrend Grubb, Ann Hamer, Kenny Hamer, Kathy Hamilton, Dwight Iler, Flo Jones, Michael Joyce, Beverly Kline, Melissa Knapfel, Robert Laufert, Addie Lewis, Betty Lewis, Susie Lewis, Mary Livingston, Nancy Lutz, Mandy Martin, Marie Hilton Mason, Everett Miller, Sandy Miller, Virginia Miller, Yones Mossa, Pat Nahs, Gale Nwaba, Marcella Miller Pak, Flo Patterson, Pep Peplinsky, Thelma Peplinsky, Cindy Peterka, Roy Phelps, Diane Pinter, Valdez Preston, Mary Pyles, John Reed, Steve Rehak III, Marlene Rydzewski, Rayvanna Scheper,

Steve Scott, Len Shindell, Rose Shockey, Ellen Smith, Janice Smith, Courtney Speed, Lucy Thornton-Berry, Sharon Toll, Helen Topper, Josephine Toro, Stan Trainum, Pat Van Tassel, Paul Willis, David Wilson, Dottie Wisniewski, Sue Wright, Rich Writkowski, and Carolyn Miller Zakwieia.

An account of the transformation taking place in the Dundalk and Turner Station steelmaking communities began in a National Endowment for the Humanities seminar taught by Sidney Mintz, and evolved into a dissertation in the American Studies Department at the University of Maryland, directed by John Caughey, Hasia Diner, Myron Lounsbury, Claire Moses, and Mary Sies. Linda Shopes shared her extensive understanding of Baltimore's industrial history, and encouraged me at every stage of this project. Other scholars who contributed include Bill Barry, Suzanne Beal, Louis S. Diggs, John Hinshaw, Laurie Mercier, Christopher Niedt, Shirley Parry, Kathy Peiss, Jo-Ann Pilardi, Mark Reutter, Cecilia Rio, Jo Ann Robinson, and Linda Zeidman.

The Dundalk-Patapsco Neck Historical Society and Museum and the Turner Station Heritage Foundation generously shared the rich resources in their archives, and Courtney Speed and Jean Walker helped me locate information and photographs. The Louis S. Diggs Collection of Black Life and Culture in Baltimore County, Maryland, and the published research conducted by Diggs and James Watson were invaluable sources of photographs and the previously neglected history of Turner Station. Participation in the Oral History Association enhanced my understanding of social history, oral history, and ethnography. Special thanks go to those who provided research assistance: Jane Willeboordse of the Dundalk Renaissance Corporation; Juliana and Homer Schamp; Mary Logan of the Metropolitan Council; Jane Harrison; Sharon Casey, Yvette Dickson, Janice Evans, Lon Fen Hou, Ellen Smith and Michelle Virgin from the Community College of Baltimore County Dundalk Campus; Jeminat Adeshina; Raymond Schamp; and others too numerous to name. Several people responded to the manuscript, including Constance Bloomfield, Patricia Ferraris, Janice Kisner, Bernadette Low, Patricia Owens, Zora Salisbury, Meredyth Santangelo, Ellen Smith, Susann Studz, and David Truscello.

The enthusiasm and clear-sighted guidance of Peter Potter, editor-in-chief at Penn State Press, and the work of his talented staff, especially Tim Holsopple, Patty Mitchell, and Jennifer Norton, brought to fruition the publication of *Wives of Steel*. Michael Frisch was exceptionally generous with his time and expertise in critiquing the manuscript. The support of Betsy Gooden, Frank Greene, Anne Griffith, Susan Laugen, and Patricia Paluzzi sustained me through the process. Christopher Whitman provided the best of his music, humor, and perspective—thank you.

Introduction

> My husband has always been the breadwinner. He made twenty dollars an hour as a steelworker at Sparrows Point, and in twenty years we'd never had a problem. Eight months ago my husband came home and said that everybody had to cut their days back to two or three days a week. I knew I was going to have to go to work full-time.
>
> —PATRICIA GIORDANO, SEPTEMBER 1988

In the spring of 2003 a *Baltimore Sun* headline read, "The Rise and Fall of Life and Steel at the 'Point.'" The accompanying article went on to describe the end of a tradition of generations of steelworkers earning wages at the Sparrows Point, Maryland, steel mill that propelled their families into the middle class. Journalistic accounts of deindustrialization in the United States have consistently used the "rise and fall" analytical framework. They follow the growth of the steel industry through the first half of the twentieth century and then document its steady, disastrous decline beginning in the 1970s. They emphasize the Golden Age of steelmaking in the 1950s and 1960s and bemoan the loss of high-income industrial jobs through the last three decades of the twentieth century. What emerges is a stereotypically tragic view of the ways in which deindustrialization has resulted in the loss of a world of steelworker prosperity.[1]

When the focus is on Sparrows Point in the 1950s and 1960s, on the fortunes of white men who earned high incomes, and on black men who were about to claim the legal right of access to better jobs, the story of deindustrialization appears as a calamitous collapse of the good life in steelmaking communities. This story, however, obscures the larger picture. Between 1887 and 2003 there have been only three decades—the 1940s, 1950s, and 1960s—when Sparrows Point steelworkers had both union wages and abundant jobs. Even during those three decades, African American men were, with few exceptions, limited to lower-paying laboring jobs, and women were excluded from virtually all production jobs, except during World War II. For more than a century, the neighborhoods surrounding the Point participated not in a monolithic prosperity but in a history of complex social relations—between women and men, between blacks and whites, and between skilled and unskilled workers.[2]

Sparrows Point, located on a peninsula southeast of Baltimore, is the name of the steel complex built in the 1880s and owned for 87 years by the Bethlehem Steel Company. It is also the name of the original company town, the planned community built to house white and black steelworkers and their families between 1887 and 1974. Isolated and self-contained between 1887 and World War I, the first decades of the town of Sparrows Point were characterized by a set of social relations different from the working-class experience of that time in the city of Baltimore. Two-thirds of the men working in the mill in 1900 were native-born white men, of whom 92 percent were from rural Pennsylvania, not from Maryland or further south. One-third of the paid workforce at the Point was African American. (This stands in significant contrast to the steel mills in New York, Ohio, and Pennsylvania, which hired Eastern European immigrants—but few African Americans—for unskilled laboring jobs prior to World War I.) Though the environment was rigidly segregated, it still provided more economic and social opportunities for African Americans than agricultural towns in the Upper South did, and a cohesive black community formed and flourished. The town of Sparrows Point also had a distinctively young population, composed predominantly of men and women in their twenties and thirties. These workers found it relatively easy to move themselves and their small families to a new location. The Sparrows Point steel mill employed few of the European immigrants that were crowding into Baltimore and other American commercial and industrial cities. For all of these reasons, the company town had a different ethnic composition than the city.[3]

Adjacent to the Sparrows Point steel mill are two other communities—Turner Station and Dundalk. Dundalk, just a few miles from the steel mill, also began as a planned community, and it was built under the sponsorship of the Bethlehem Steel Company to house the abundance of white workers employed at the Point during and after World War I. Turner Station is an African American community that grew more organically, beginning late in the 1800s as a small black settlement and continuing its growth as black steelworkers from the Point built homes there or settled into one of the housing developments built as part of the wartime effort in the 1940s.[4]

Wives of Steel is the story of 115 years of steelmaking in these locales—but told from a gendered perspective. In the many fine studies of steelmaking communities, gender has not been the central focus, and women often are relegated to a single chapter or a courteous mention. Usually, it is assumed that America's industrial history is the story of men putting in long hours at backbreaking work, men struggling against imperious company owners, and men organizing powerful unions to protect their wages and working conditions.[5]

But in Sparrows Point, Turner Station, and Dundalk, gendered social worlds evolved that were characterized by the sharp segregation of men's work in a male-dominated steel mill from women's work in households organized around shift work schedules. It is the role assumed by large numbers of steelworkers' wives that begs our attention—and has been underestimated both in historical renderings and in the construction of contemporary accounts of deindustrialization.[6] As early as the 1880s, women in the company town of Sparrows Point played critical economic roles. They prepared food and did laundry for children and for husbands and adult sons who were working in the mill, but also cooked and cleaned for the unmarried male steelworkers they took in as boarders and roomers to add essential income to the household.

Women went to work at the Point in the tin mill just before World War I, filled the paid workforce shortage during World War II, and took temporary jobs in banks, restaurants, department stores, and offices during periods of long strikes. Wives of steelworkers resumed the role of economic contributors after 1970 as thousands of predominantly male industrial jobs vanished from the United States. Only briefly, in the 1950s and 1960s—a period of affluence for steelworkers, because the United Steelworkers of America (USWA) protected good wages and sparked high productivity in the American steel industry—did many wives of Sparrows Point steelworkers assume the role of full-time, lifelong housewives and consumers dependent on the income of steelworkers who were their families' sole breadwinners. That Golden Age was wonderfully prosperous but brief.[7]

The period between World War II and 1970 is, in fact, an anomaly in many ways. It certainly was the only era during which so many steelworkers earned enough to support a family, allowing wives to live relatively comfortably as homemakers with no need to bring additional income into their households. Those years of prosperity have largely obscured the fact that prior to World War II many steelmaking families relied on income brought into the household by wives. *Wives of Steel* corrects this misconception by recounting the unpaid work done by the wives of Sparrows Point steelworkers before World War II and by highlighting the importance of income earned in the paid workforce by contemporary women.[8]

Wives of Steel is not a journalistic account of the current crisis facing America's steelmaking communities. Rather, it is an ethnographic study based on more than eighty formal interviews conducted over a fifteen-year period. It uses the voices of women, as well as several men, to gain a unique perspective on the ways in which the Sparrows Point mill has influenced personal, family, and social experiences in the neighborhoods surrounding it. This ethnography is placed within a historical structure that documents the founding of the mill and company town, the expansion of workers' communities during two world wars, the emergence of steelworker prosperity in the post–World War II era, and the decline of jobs in steel after 1970.[9] The book investigates the gender dynamics in black and white steelworker communities, beginning at the end of the nineteenth century with the young steelworkers and their wives who migrated from rural areas to claim better-paying industrial jobs at the Point. It extends that gender analysis through the early twentieth-century period of rapid expansion in America's steel industry and into the late twentieth-century post-industrial era that sent steelworkers' wives into the full-time paid workforce in large numbers.[10]

Wives of Steel makes a unique contribution to the literature on steelmaking communities because it is based on the voices of people describing life in their own households and neighborhoods, including the voices of women like Gloria Robinson, who grew up with a certain rhythm in her household: "My father and his three brothers all worked at Bethlehem Steel, and my husband worked at Bethlehem Steel, and everything in the family revolved around the mill schedule." Five generations of steelworker families have struggled to organize their day-to-day existence around steelworker husbands absent from their homes because of long shifts in the mill. Tom and Theresa Porsinsky agreed that "shift work makes it a difficult life and can wreck the family. We're both prisoners of Tom's schedule . . . [and] you never get used to it, you just put up with it."[11]

Women married to steelworkers are placed squarely at the center of this study, with a particular focus on the continuity of women's economic contributions to their households. I am emphatic about the turmoil that the loss of jobs in steel has caused the people who lost them, and I listened especially to those men who can no longer support their families without help from their wives, and who, according to Doug Ingles, "felt wimpish, like they weren't holding up their end of the deal." (A former steelworker, Eric Logan, described the ethos of manliness inside the steel mill: "It is dangerous, hard, physical work. Just the word 'steel' implies toughness; look at Superman, he's the 'Man of Steel.'") This study, however, focuses sharply on the economic role of steelworkers' wives in order to correct the assumption that, from the late 1800s until the 1970s, America's steelworkers earned enough to single-handedly provide for their households.[12]

A gendered perspective challenges the standard accounts of deindustrializing American communities, which have almost universally portrayed a gloomy downward trajectory for the people who inhabit these places in a postindustrial age. *Wives of Steel* widens the perspective to include a much longer span of history in Baltimore's steelmaking communities, and it documents the ways in which steelworkers' wives provided essential economic contributions to their households during times of low wages, wars, lay-offs, and lengthy strikes. The second half of the book emphasizes the independence developed by steelworkers' wives who, after 1970, went into the paid workforce as jobs in steel declined. These women were influenced by both deindustrialization and the loss of household income, as well as the changing gender identities prompted by feminism after the 1970s. Wives of steelworkers discussed with me the ways in which they had gained personal power in the larger social world of the paid workforce, finding that, as Willa Martin said, "The younger generations of women have the benefit of association with a wider variety of people from which to draw."[13]

My research relied on archival and census sources, as well as ethnographies and oral histories that I began collecting in the mid-1980s. Rufus and Frederick Wood, the enterprising New Englanders who planned and constructed the Sparrows Point steel complex, documented their plans for the company town during its first two decades through letters and reports currently archived at the Hagley Museum and Library in Wilmington, Delaware. In addition to the papers of the Wood brothers, the manuscript census of 1900 chronicles life in the company town of Sparrows Point by delineating the composition of each household. That census lists the address for each household in the community and the name, age, birthplace, and relationship to head of household for each occupant, including family members and boarders. The census goes on to list number of years of education for each person and whether or not he or she was a naturalized citizen. For each adult woman the census gives an accounting of infant and child mortality by listing the number of children born and the number surviving. For each adult man an occupation is recorded that allows an analysis of the segmentation of the paid workforce by ethnicity and race. This record of the composition of neighborhoods in the company town by occupation, level of education, ethnicity, and citizenship status illuminates a multifaceted panorama of the people who lived and worked at Sparrows Point during its first thirteen years.[14]

The 1900 census is particularly valuable because it is the first manuscript census available that includes Sparrows Point. Most of the 1890 census, including the entries for Sparrows Point, was destroyed by fire. By 1910, the census records for the company town no longer contained all of the mill's steelworker families because a streetcar line, introduced in 1903, brought hundreds of steelworkers to the Point each day from working-class areas of Baltimore. Although the company town flourished until Bethlehem Steel began demolishing it in 1954, after 1900 increasing numbers of steelworkers commuted from surrounding areas.[15]

Because the Sparrows Point steel mill continues to produce steel for America and the world, the communities of Dundalk and Turner Station are still vibrant and remain homes to families with generations of steelworkers. Between 1985 and 2003, I collected eighty oral history and ethnographic interviews with adult women and men, white and black, all of whom had worked or were working at the Sparrows Point mill, were married to Sparrows Point steelworkers, or were longtime residents of Sparrows Point, Turner Station, or Dundalk. I have come to see this community from many different points of view, and I have been able to watch closely as residents respond to continual change. Because I had a sharp focus on gender, there are many specific details about this community that I did not study or could not include in a short book; however, as I continued my research, I found that the social dynamics of race and class were inseparable from issues of gender. The large number of interviews that I conducted over a long period of time is one of the major contributions this study makes to the literature on industrial communities.[16]

The interviews I conducted usually lasted for an hour, although some went on for as long as three hours. All of the individual interviews were audio taped, and the oral history interviews were transcribed. In some cases I conducted a series of interviews with one individual, and occasionally I interviewed married couples or small groups of people. At the conclusion of the study, I asked individuals to read what I had written, and then I conducted follow-up interviews that focused on how people interpreted what I had said about their community.[17]

During these interviews, I asked members of steelworker families to describe the role that gender played in determining both the work culture at the Point as well as the rhythms of family life in the community. I asked them about race, class, and other hierarchical systems that existed in the mill or in their neighborhoods. In particular, I asked how deindustrialization had affected gender roles and the organization of families as jobs at the mill declined. I also asked how race relations at the mill and in communities were experienced and perceived.[18]

In analyzing this collection of interviews, I looked for patterns that would provide insights into the configuration of gender within families in the area. I found a complex interwoven set of patterns. Gender loomed large, but it intersected with race and age, as well as with the social relations of class in the community, which was closely connected to status in the mill hierarchy. Within a community where people were linked by a common relationship to a dominant industry, there were sharp contrasts in the perspectives on the Sparrows Point experience. Even interviews with a single person always yielded a complex, nuanced set of reflections on the hardships and rewards of living and working in a steelmaking community.[19]

For researchers using ethnographic and oral history interviews, it is clear that historical memory and renderings of present circumstances are socially constructed; the narratives people tell are framed by their social experiences. Oral historians have emphasized the ways in which subjects being interviewed filter their lives through their perspectives on dominant institutions. Ethnographers have emphasized the relationship between the autobiographical particulars of interviewers and the responses of their subjects. People who are being interviewed respond to questions by telling their narratives of life events from a particular point of view that reflects social status, belief systems, occupation, gender, race, and myriad other particulars. In addition, an interviewer brings an identity to each interview exchange that elicits not just answers to her questions, but also responses to her social self.[20]

One pattern emerged most distinctly during seventeen years of interviewing black and white men and women of varying ages. That pattern lies in the differences in the perspectives of the men, and some women, who have worked at the Point, and in the perspectives of the wives of steelworkers, some of whom have never been inside the Sparrows Point mill. Men who were working at the Point at the time of their interviews were matter-of-fact about the work they did, and they often would begin their descriptions of their jobs in a self-denigrating manner by saying, "It pays the bills," but then go on to describe accomplishments, skills, and friendships that were the source of pride. I started interviewing men who were working at Sparrows Point in 1988, at the height of layoffs in the steel industry and after Bethlehem Steel had been reducing its paid workforce for more than a decade. Most of the men I talked to described themselves as working in an industry where fear and resentment, caused by the possibility of job loss, were widespread. Unprofitable departments in the mill were being closed, and new union contracts typically contained cuts in pay or benefits. On May 7, 2003, after a long period of financial difficulties, the Sparrows Point steel complex, along with six other Bethlehem Steel plants, was sold to International Steel Group Inc. of Cleveland, which is salvaging large steel companies in an effort to combine them into a company that can survive in the highly competitive global economics of modern steel production. Jobs and even the health benefits of retirees are vanishing from a community built upon the essence of the American Dream—hard work rewarded with economic security.[21]

Although my primary focus is on gender relations, it was clear from the outset of this project that race and class were inextricably interwoven with the gender dynamics within steelworker families. From its beginnings in the late 1880s, the Sparrows Point steel mill actively recruited African American men, restricting them to the lowest-paying laboring jobs, which impacted black families and the economic roles of wives in those families. A clear pattern of distinctly different ways of seeing race relations emerged in the series of interviews conducted in the 1980s and early 1990s: white steelworkers insisted that everyone was treated the same, "as long as you're willing to work," while black steelworkers remember the plant's history of discrimination as a painful reality. In the follow-up interviews that I conducted in 2002, black and white men referred to major changes at the Point. Interracial teamwork, and occasionally even friendship, was forged in ways that did not seem possible for earlier generations of steelworkers.

Because the company town of Sparrows Point was demolished thirty years ago, interviews with "Sparrows Pointers" require a special context because they reach back across a quarter of a century. Families that had lived for several generations in the town of Sparrows Point held tightly to memories soaked with emotion and recalled stability, friendliness, and employment security in a cohesive, well-ordered company town that was antithetical to the neighborhoods across the state of Maryland, where aging Sparrows Pointers live today.[22]

Diverse perspectives are embedded in the memories of the African American section of Sparrows Point, including some bitter reflections on the employment of black women as domestic help in white homes. However, even though the black section of Sparrows Point was deliberately planned and maintained as a separate, segregated neighborhood, the African American families I interviewed remembered their community fondly. African American men who endured their restriction to laboring jobs in the steel mill spoke of the comfort of going home to a quiet neighborhood of well-kept homes and carefully tended gardens in a community free of overt racial conflict.[23]

Memories of the company town are part of a contemporary social experience that includes reunions held regularly by Sparrows Pointers. These reunions are festive, and the social discourse has an idyllic ring to it. It is only with considerable probing that Sparrows Pointers remember the more complicated aspects of the company town. I spent several hours listening to a group of white men and women talk about the special pleasures of living in Sparrows Point. Fatal accidents were never mentioned. Unions were never mentioned. Race was never mentioned. It was only after I asked about the hiring practices of boardinghouses that someone said, "Well, of course, all of our mothers had black women come in and iron."[24]

The men and women who did not have such positive memories of the town of Sparrows Point were those who had been active in the union movement of the 1930s. They remembered the determination with which Bethlehem Steel had banned union organizing in the town of Sparrows Point, and reminded each other that the alliances that were built among men working at the Sparrows Point mill crossed racial boundaries, not the boundaries of the company town. African Americans who commuted to Bethlehem Steel from West Baltimore joined native-born whites and European immigrants in the campaign for a union. African Americans who lived in the town of Sparrows Point, like their white counterparts, could not aid the union movement without risking eviction from their homes and the loss of their jobs.[25]

Agreeable memories of safe and close-knit communities also reverberated in the conversations of blacks and whites reflecting on their lives in Dundalk and Turner Station between 1950 and 1970, which to their way of thinking were more predictable, less turbulent times. Jacinta Cole told me about Turner Station, "I don't think there is any other place in the world where I could have been more comfortable growing up than in Turner Station," and went on to describe churches that took care of the elderly and teachers in their segregated schools who insisted that African American children graduate with pride and skills. Reluctant to tell a white interviewer about hurtful experiences in Dundalk, it was only after I acknowledged my own awareness of racial tensions that Cole expressed her gratitude that, "we were sheltered by our parents from the rejection in Dundalk."

In interviews with and about women married to steelworkers, diametrically opposed themes recurred. The first theme is unequivocal admiration for the wives of Sparrows Point steelworkers because of their hard work. Especially when daughters and granddaughters remembered the era prior to World War II they expressed respect, even awe, at the sheer enormity of the work done by steelworkers' wives to maintain households, accommodate boarders, raise families, and contribute as wartime workers. The second theme acknowledges the extent to which wives were constrained and at times protected by the social organization of their steelmaking community, and wives of steelworkers were sometimes even portrayed as trapped in drudgery and living in a very small world. In some cases, daughters and granddaughters depicted the post–World War II generation of steelworkers' wives as capitulating to husbands demanding excessive attention and services. Often steelworkers' wives would reveal how they experienced the two themes in their own lives: the pride in self-sacrificing attention to their husbands and families, along with resentment that unpaid, and sometimes unappreciated, work had dominated their lives.[26]

Deindustrialization sparked the pivotal change in the gender dynamics within steelworker families. Men who had worked at the Point while the mill was prospering described the period before 1970 as a good time and expressed pride in their roles as steelworkers and as the providers of family income. Individuals who have personally experienced the loss of jobs in steel presented more complicated points of view. Women who found satisfying jobs outside the home admitted that they were torn between the loss of a certain kind of family unit and the rewards of their own increased independence and personal autonomy. Men and women in the throes of losing jobs and stable social identities focused on the disturbing economic changes that clouded the way they remembered the past and that dominated their horizons.[27]

In every interview people stated emphatically that their communities have been severely impacted by the decline of employment opportunities at the Point. It was the multifaceted perspectives of those interviews that allowed me to understand that gender, race, class, and age were interrelated factors that determined how differently members of steelworker families had experienced, remembered, and represented the process of deindustrialization.[28]

Wives of Steel begins by looking at the history of the three communities that accommodated the families of men who have worked at the Sparrows Point steel mill. During the period between 1887 and the 1920s, the construction of gender relations is particularly evident because the town of Sparrows Point had an overwhelmingly male population, and men were working twelve- to fourteen-hour shifts in the mill while women contributed to the business of making steel by taking in unmarried steelworkers as boarders, raising their own large families, and keeping the mill's red dust residue at bay.

Chapter 2 looks at how a male work culture was set in place inside the Sparrows Point steel mill, which has always had an overwhelmingly male paid workforce, and where women were initially forbidden entrance, even to deliver a husband's dinner pail. Although a small number of women began working in the tin mill early in the twentieth century, in their interviews both male and female steelworkers repeatedly describe the Sparrows Point mill by saying, "It's a man's world."

Chapter 3 explores the responsibilities carried by women married to steelworkers between 1887 and World War II and living either in Sparrows Point or, after World War I, in Dundalk or Turner Station. The astounding number of Sparrows Point households that took in boarders reveals that, at the turn of the twentieth century, Sparrows Point wives ran what essentially were small businesses that accommodated the large numbers of single men living in the community during the early years of the mill.

In marked contrast, Chapter 4 explores the ways in which Dundalk and Turner Station became middle-class steelworker communities during the post–World War II era. The role of the steelworker's wife was transformed, not only because she no longer took in boarders, in part because more housing was available, but also because her family no longer needed the additional income once unionization brought higher wages. After World War II, women married to Sparrows Point steelworkers had more conveniences, fewer children than earlier generations, and an income that allowed for luxuries. However, for the first time since the founding of the Sparrows Point plant, most steelworker wives lacked an economic role that included contributing income to their households, and many saw themselves as "prisoners of the swing shift."

Chapter 5 includes the voices of those women from Dundalk and Turner Station who did make attempts to work at the Point, beginning with the all-female tin flopping department, examining the few hundred women who worked at the Point during World War II, and including contemporary women who gained permanent production jobs at the Point after the 1974 Consent Decree banned sex and race discrimination. Despite reporting modest successes in improving safety conditions and in enforcing prohibitions against sexual harassment, the presence of women steelworkers in the mill has never produced pervasive or fundamental change in gender relations at the Point or in the community.

Chapter 6 explores the ways in which the decline in steelmaking jobs has pushed families in the communities surrounding the Sparrows Point steel mill to pursue a variety of alternative economic strategies. In my interviews I asked women and men to reflect on the dramatic decline in employment in the steel industry that began in the 1970s, and to describe the ways in which their own lives as well as the dynamics within their families had changed. When the economic transformations within steelworker families were examined as a change in family gender dynamics, I found something much more complicated than a straightforward picture of loss of jobs and income. When a steelworker faced the loss of his job at Sparrows Point, his entire family went through a process of reacting to changing economic conditions, and, while responses varied considerably, in many families the loss of steelmaking jobs meant that wives of steelworkers needed to contribute substantial income to the household, just as wives of steelworkers had done prior to World War II.

Chapter 7 looks at those women for whom the decline of steel was an impetus to enter the full-time paid workforce, opening up possibilities for altered relationships between women and men. In one interview after another, women in Dundalk and Turner Station described the process of negotiating for themselves a fundamental change in lifestyle, from the breadwinner/homemaker families that accommodated the long hours and swing shifts at the Point, to families in which divorced or single women are themselves breadwinners or to families in which women are sharing the breadwinner role with their spouses. Within individual families, women and men described complex and sometimes minute transformations that occurred as the wives of steelworkers, finding it necessary to contribute to their household's income, began the process of renegotiating gender relations that are characterized less by breadwinner/homemaker hierarchies and more by partnerships. Those women whose marriages ended in divorce described their own processes of gaining independence.

Finally, Chapter 8 focuses on the ways in which race and class relations intersect with the negotiation of new gender relations. Looking at changing class and race demographics in the Dundalk and Turner Station communities, it follows both black and white women as they commute to work in parts of the Baltimore metropolitan area that are outside the Sparrows Point steelmaking communities. It includes the voices of black women who found some advantages in working in places far removed from the racial hierarchy that existed in neighborhoods surrounding the steel mill, as well as the voices of white women like Marcella Knowles, who was shocked by the class biases of coworkers who "talk about my neighbors as the 'lower class' and call them 'grits.'" Both white and black women talk about their commitment to preserving the benefits of a cohesive community that is the heritage of the Sparrows Point steelmaking area, while embracing new sets of race and gender relations that encompass a larger "circle of neighbors."

Wives of Steel is not a business or labor history of the Sparrows Point steel complex; rather, it is a gendered analysis of the area's history that listens to voices of women and men as they describe the ways in which the male-dominated work culture of a steel mill affects all areas of family and community life. Although my original intention was to limit my study to the lives of women in a steelmaking community, I found that gender is inextricably linked to race, to class, and to generational differences. Between 1985 and 2003, my interviews coincided with the process of job losses that left families seeking strategies for their economic survival. The voices from a steelmaking community in the midst of a shrinking number of steel jobs uncovered an often sad but always brave and resourceful response to deindustrialization. In the communities surrounding the Sparrows Point steel mill, the strategies used by families in the face of deindustrialization precipitated a renegotiation of the roles of women and men that has been varied, complex, and far-reaching.

1

Sparrows Point, Turner Station, and Dundalk: The History of the Mill's Communities

People who lived in Sparrows Point never left their community, mainly because everything you needed was already there. You had a grocery store, you had a jewelry store, a tobacco shop, and a furniture store. Now this was all within one building. You also had a gas station and across the street you had a pharmacy. Sparrows Point also had its own bakery and we had a bakery truck, Mr. Green's bakery truck, and it was horse-drawn. He'd come around every day and he'd ring one of those old-time heavy bells.

—HELEN REED, AUGUST 2002

In 1870, American steel was in its infancy, using crude methods and producing one-sixth the quantity of pig iron as that produced in England. But the Bessemer process and the development of integrated steel mills such as Andrew Carnegie's Thompson Works catapulted American steelmaking to its eventual position of number one in the world. The Sparrows Point steel mill, built during the peak of the growth and development of America's steel industry, was an integral part of this industrial process.[1]

In May 1886, the Pennsylvania Steel Company of Steelton, Pennsylvania, purchased 1,221 acres of land on the Patapsco Neck southeast of Baltimore, Maryland. By August 1887, construction of the blast furnaces on Sparrows Point began, and between sixty and seventy homes for steelworkers and their families were built. The land, the steel mill, and the company town all derived the name Sparrows Point after Solomon Sparrow, who had built his estate, "Sparrow's Nest," in the area in 1664. On October 23, 1889, the first pig iron was cast at Sparrows Point, and in 1891 the rapidly expanding steel complex was incorporated as the Maryland Steel Company.[2]

Between 1887 and 1900, the town of Sparrows Point expanded as representatives of the Maryland Steel Company recruited men to work in the mill. The manuscript census for the year 1900 documents 3,520 people living in a community distinctive by virtue of the fact that two-thirds of the adult citizens were married or single steelworkers, and one-third were wives of steelworkers. Originally intended as a community of families, the large number of single men hired to fill jobs in steel required a system of boarding that was managed, by and large, by steelworkers' wives. Those 434 wives did virtually all of the cleaning, cooking, budgeting, shopping, and laundry for the two thousand men employed in the mill.[3]

The census reveals a complex social structure of families and single men, and the dates and places of birth of the adults in the town provide a directory to the migration patterns. The overwhelming majority of steelworkers who had come to Sparrows Point by 1900 were native-born men who left nearby rural areas to take advantage of the promise of steady work and the relatively higher wages of industrial labor. Most of the white workers who moved to Sparrows Point came, either alone or with their families, from Pennsylvania and Maryland. About two hundred native-born white American men appearing in the 1900 census came from further away, including New Jersey, New York, and even two from Colorado.[4]

The Sparrows Point steel mill relied more on African American men for its paid workforce than it did on immigrant labor, a characteristic unique to the steel mills in Baltimore, Maryland, and in Birmingham, Alabama, because of their location in southern states. In the 1880s and 1890s there was an ample supply of African American men eager to leave jobs as sharecroppers, tenant farmers, and farm laborers in rural Maryland and Virginia for the higher paying industrial jobs at Sparrows Point. African Americans at Sparrows Point in 1900 were 31 percent of the labor force, in contrast to the steel mills of this era built in Pennsylvania, Ohio, and New York, where large numbers of Eastern European immigrants but few African Americans were hired prior to World War I.[5]

In 1900, there were 58 married couples, almost always with children and boarders, living in the African American section of Sparrows Point. Most of these black families renting houses in the company town and sending men

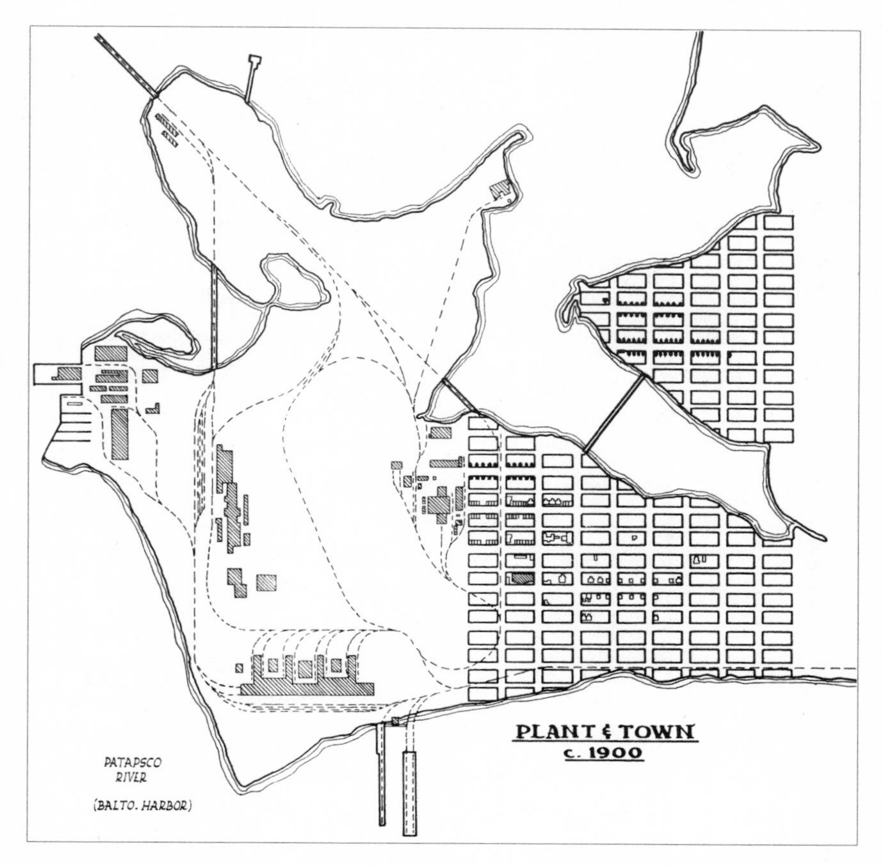

FIG. 1 New Englanders Frederick and Rufus Wood planned and developed the steel mill and workers' housing at Sparrows Point, represented here in a plan drawn in 1900.

into the steel mill were from rural Maryland and Virginia. Single black men boarding with black families were more likely to have migrated to the Sparrows Point mill from North Carolina, and a few came from as far away as South Carolina, Tennessee, Georgia, and Missouri.[6]

European immigrants appear in the 1900 Manuscript Census for Sparrows Point, but their numbers are small. Men with families were recruited from England, Scotland, and Wales because of their essential skills as machinists and drill press operators. Sixteen German families appear in the census; the men worked as laborers while their wives took in boarders and raised children. The Irish were the largest immigrant group at Sparrows Point in 1900, but they numbered only 104 workers. There was one household of Russian immigrants, and five single Russian men living in the shanties, as well as one single man from each of the countries of Austria, France, Hungary, Norway, Poland, and Switzerland. Two men, one black and one white, came to Sparrows Point from Cuba. Prior to World War I, the reliance on immigrant labor at the Point was minimal. Like all other steel enterprises of this time, no women were on the company's roster except as clerks in the company store. A few unmarried women were listed in the 1900 census as cooks, servants, and seamstresses.[7]

The social structure of the company town of Sparrows Point constituted a social hierarchy based on class, ethnicity, immigrant status, race, and gender that would dominate this steelmaking community for the next century. At the end of the nineteenth century, class and race differences were expressed in the blunt language characterized by a statement printed in 1892 in a local paper, the *Union*:

> The men engaged here are of all grades and classes, ranging from the intellect required for the construction and conducting such an immense plant and business, down to the man who only knows enough to draw his money on payday and get drunk. Many of the laborers are negro [*sic*], Russians and Hungarians and there are some who don't know their nationality. Among the better classes are many Americans, Germans, and Irish.

Like other steel mills of the time, recruitment for workers at Sparrows Point relied on a hierarchical arrangement, and in the communities surrounding the mill the ethic by which social relations were structured in 1900 continued to influence the ways in which people thought about work, neighborhood, and family life.[8]

Two New England brothers, Frederick and Rufus Wood, designed the company town of Sparrows Point to accommodate the workforce for their plant of the same name, and the Pennsylvania Steel Company began construction on the complex in 1887. The collaboration of the Wood brothers is well documented by correspondence between the two. Rufus moved to Maryland to direct the construction of the mill, while Frederick remained in Steelton, Pennsylvania, where he managed another major steel facility for Pennsylvania Steel. Rufus wrote to Frederick frequently in an attempt to apprise his brother of the progress made and the problems encountered at the new steel works in Maryland. These letters reveal the goals that Rufus had for building an efficient mill and at the same time recruiting a workforce that would fit the need for managers, skilled craftsmen, laborers, mer-

FIG. 2 The Sparrows Point steel mill produced its first heat of Bessemer steel at 4:17 P.M. on August 1, 1891. By 1900 the mill had attracted more than 3,000 residents, most of them migrating from rural Maryland, Virginia, and Pennsylvania in search of industrial wages.

chants, ministers, and physicians. Sparrows Point was to be a completely self-sufficient town, fulfilling the workforce needs of the steel mill as well as the commercial, recreational, educational, and spiritual needs of the citizens.[9]

Foremost in the planning of Sparrows Point were the principles of efficiency and order, an innovative concept aimed at maximizing profits, and one to which Rufus assigned four major corollaries. First, the residential design of the company town was intended to accommodate a workforce that would be segmented by skill, ethnicity, regional origins, race, and gender. Second, using what they believed to be a highly enlightened system of community management, the Wood brothers instituted an educational curriculum that instructed students in those skills that would best serve the needs of the steel mill. Third, the engineers of this planned community imposed legal restrictions on the consumption of alcohol within the environs of Sparrows Point, intending that this restriction, along with the company-sponsored churches, would ensure sobriety among their overworked steelworkers. Finally, women, though not permitted in the works, were assigned an essential role at Sparrows Point as the providers of domestic services to men whose hours outside the mill were in many cases limited to eating and sleeping.[10]

Class and race divisions were explicitly delineated in this planned industrial community by a hierarchical arrangement of east-west streets lettered from A through K with four major divisions separating the socially significant groups. The lot on A Street had been reserved for the family of Frederick Wood, but it was never used because he moved his large family to Lutherville, a suburb north of Baltimore. Located on B and C streets were the large frame homes of the mill superintendents, the principal of schools, the company physician, and General Manager Rufus Wood. These men were the elite of the town and their elegant homes were cared for by personal servants and surrounded by expansive lots. On D Street, churches, schools, and stores were built, and that street served as the first divide between social classes.[11]

North of D Street were located the homes of foremen and skilled white workers in an area that accommodated 2,189 men, women, and children. This was an area of red brick row houses with small front porches, a kitchen annex, six rooms, and a back yard. The company proudly described these homes as "substantially built, having baths supplied with hot and cold water, inside closets and sanitary plumbing, as well as ranges and kitchen equipment." Houses in this section of Sparrows Point were small—just 13 feet by 28 feet—and crowded, because on average seven or eight people lived in each household, including one or two boarders.[12]

The white community of Sparrows Point ended at Humphrey's Creek, a natural barrier that would be used by the town's architects for the purposes of racial segregation. An 838-foot walkway extended across Humphrey's Creek onto the stretch of land that contained Sparrows Point's black housing. In keeping with the social hierarchy of the community, houses in this section were smaller and lacked the amenities of running water and indoor plumbing. In order to compensate for the lower wages earned by black men relegated to laboring jobs, wives in Sparrows Point's African American community increased household earnings by taking in larger numbers of boarders than was common in the white community. In this densely populated black community, it was more common for extended family members to join the household, and on average eleven or twelve people, including boarders, lived in each home.[13]

The deliberate segregation of black laborers "across the creek" was con-

FIG. 3 In the newly constructed town of Sparrows Point, the modest but sturdy homes of white workers were considerably smaller and less lavish than those of the managers but larger than those in the African American section of town.

trary to the racial patterns of Baltimore in the 1880s. After the Civil War and until 1911 black citizens lived in all twenty wards of Baltimore. In the central, southern, and eastern sections where the majority of the black population resided, all black and all white blocks lay interwoven with blocks that were racially mixed. It was during the 1880s, when race relations in Baltimore were most fluid and permissive, that blacks began moving into Northwest Baltimore, a predominantly white residential neighborhood of comfortable dwellings. Although the Northwest neighborhood would eventually become a segregated African American community, the practice of residential racial separation was not firmly in place in Baltimore until two-and-a-half decades after the community of Sparrows Point was established.[14]

The Sparrows Point design, therefore, was not simply an effort on the part of the town's architects to conform to the southern racial practices of Baltimore. Ironically, it was two New Englanders, the Wood brothers, who imposed segregation on the industrial town that they built in the southern state of Maryland. They had determined that Maryland was a region with a disciplined and desirable African American labor force, but a region that also had the potential for chronic racial tensions dating back to the state's slave history. Residential segregation at Sparrows Point was a distinctive engineering approach to prevent racial problems in a strategically planned workers' community.[15]

Within two decades, the white leadership of Baltimore had taken favorable note of residential segregation in the town of Sparrows Point. In 1906 an editorial in the *Baltimore Sun* touted the company town as an exemplary system of racial separation: "The race problem it has practically solved by putting the blacks on a far side of broad waters of a creek apart from the whites. Negroes form one-quarter of the population and yet the relations between black and white are quite pleasant."[16] The assessment that "relations between black and white are quite pleasant" is difficult to document. There is reason to speculate, however, that the men, women, and children—both black and white—who migrated from small rural towns in Maryland, Pennsylvania, and Virginia to work at Maryland Steel did find Sparrows Point an appealing community. Particularly after the economic depression of the 1890s, agricultural communities in the Mid-Atlantic region offered little more than low wages and a stagnant economy. At Sparrows Point, a company store and attractive housing that was maintained by the company were among the considerable benefits, along with industrial wages, that drew individuals and families eager for economic and social improvement.[17]

The safety, order, and access to decent housing that Sparrows Point offered must have had particular allure for young black men and families eager for economic advancement and desperate for freedom from the most extreme forms of racial oppression in small towns in the turn-of-the-century South. Residential segregation at Sparrows Point had both positive and negative consequences for African American families. Residential segregation meant smaller, more cramped living spaces, and it went hand-in-hand with occupational discrimination that denied black men access to skilled jobs in the mill. But the clear social boundaries in the town, combined with a company police force that curtailed violence, provided order and minimized overt racial conflict. Residential segregation also encouraged the development of a strong and resourceful black community through the growth of black institutions, especially schools and churches.[18]

Despite the fact that the Wood brothers patterned the company town of Sparrows Point on their hometown of Lowell, Massachusetts, the center for an American textile industry that in the 1820s and 1830s employed mainly female operatives, no women worked in the Sparrows Point steel mill. Women fit into the workforce of the Sparrows Point industrial system by

FIG. 4 The Sparrows Point Company Store, shown here around 1890, provided access to an array of consumer goods that could be purchased on a credit system that deducted the cost from a steelworker's pay.

serving as a devoted crew of laborers in domestic service conducted mostly in their own homes. Most households in Sparrows Point in 1900 took in at least one boarder in order to help with the rent. This meant that the wives of both white and black steelworkers were taking in boarders in order to increase the household income. Despite residential racial segregation and the restriction of black steelworkers to low-paying laboring jobs, black and white wives in the company town shared the common experience of cooking and washing for boarders as well as their own family members.[19]

Rufus Wood detailed a plan for workers' education that assumed that married couples would locate in Sparrows Point, but in 1900 family households were a minority of the town's population. Of the nearly 2,000 black and white men living in Sparrows Point in 1900, only about 28 percent headed their own households. Married couples numbered only 434, while the number of single men living in the town was almost three times that number.

Through the first half of the twentieth century, the company town of Sparrows Point grew into a community where men could walk to work, wives could shop and raise children with conveniences that were unusual for that era, and families of skilled white steelworkers had an opportunity for upward mobility. Crystal Young, who was born and raised in Sparrows Point, recalled the housing benefits that made the company town so much more desirable than the more expensive housing in Baltimore:

> My mother and father lived in what they called the bungalows, which was a one-floor home. It was down near the water heading towards Jones's Creek. From there they moved to H Street, where the homes were larger. By the time the last two children were born we were living on E Street. We had a nine-room house and paid $17-a-month rent. Anything that had to be repaired, painted, wall-papered, was all done free by Bethlehem Steel, every bit of it. If there was anything wrong you called the real estate office. And if your whole house needed wallpapering they would move you into another house and completely do your house and then move you back in.

Whether in the white or the black section of Sparrows Point, having the convenience of a well maintained home was considered a privilege, though in the African American section the rough plastering made wallpapering impossible.[20]

The most miserable living arrangements that existed on Sparrows Point consisted of 337 shanties that were located directly beside the steel furnaces and that housed single male laborers, primarily immigrants and blacks. The shanties were small wooden structures, the space of which was almost completely filled with four bunk beds, allowing shanty residents room to do little more than dress, undress, and sleep. The population of the shanties fluctuated according to the labor needs of the mill, so these steelworkers were the most transient, had the least job security, and enjoyed the fewest benefits from the amenities of the town. The complexity of the social hierarchy at the Point is exemplified by the ethnic composition of the shanties, where both African American and white men were housed and where, in 1900, two-thirds of the white men were Eastern European immigrants. Steelworkers living in the shanties occupied a social space further removed from family life and bereft of the domestic services of Sparrows Point wives. The ranks of men who occupied the bunk beds in the crowded shanties swelled and shrank with the company's need for laborers. These itinerant men did not eat with families or in restaurants, but in mess halls run by men who worked for the company.[21]

The town of Sparrows Point included a school for white students in 1888, and in 1903 a new brick building, Sparrows Point Elementary School, was built to accommodate 451 white students. Black students attended grades one through eight in a frame structure, which was gradually enlarged. High school education for black students remained unavailable in Sparrows Point until 1939 when, after several law suits by African American parents, Baltimore County was pushed to open three high schools for black students, including the George F. Bragg High School in Sparrows Point.[22]

The town offered other benefits as well. Recreational facilities, which were also segregated, included a restaurant, a bathing beach, and the Smoking Hall for unmarried men who were boarding in the town. Seven churches were built, including Methodist, Presbyterian, Episcopal, Lutheran, and Catholic churches for white congregants, and a Methodist Episcopal church for black congregants. In 1893 African American steelworkers organized their own Union Baptist Church.[23]

In 1900, the company store provided easy access to the essentials of food and clothing for Sparrows Point wives who did the budgeting and shopping. By the 1920s, the store had expanded into a "department" store that sold clothing, furniture, and food, including cattle, sheep, hogs, and chickens that were raised on the Point. A dairy delivered fresh milk directly to steelworker wives who fed large households, which included boarding single men. The company store found various ways to conform to Jim Crow practices, including segregating black costumers to separate sections of the stores and having African Americans pick up their purchases at a window in the rear of the store.

Sparrows Point wives also were helped with childrearing. The company store included children's clothing and a seasonal supply of toys that served as recreation for white children who were able to wander through the company store on their own:

> When I came to Sparrows Point from the Eastern Shore I was twelve years old, and I went to the company store and saw this Betsy-Wetsy doll that I wanted so badly for Christmas. The company store never had a big assortment of toys except once a year. They only brought them in at Christmas time and there was an upstairs, but it was roped off, and the day after Thanksgiving they'd

> cut the rope and, oh, my God, every kid in the Point was down there.

A century of translating good wages at the Point with access to consumer necessities and even luxuries began at the Sparrows Point company store.[24]

Wives who ran complex households could let children spend their free time browsing in the company store, and for families that had sufficient income, children could buy from the company store without having a penny in their pockets: "We didn't use cash in the company's stores. They had coupon books and you were shopping on credit, but it was coming out of your dad's pay. Our father, Ed, had a charge number, GH7, so us kids would go in and anything we wanted we would say, 'Ed Arnold, GH7.' Then on payday they would take it out." Sparrows Point wives were relieved of considerable burden because their children could safely stroll around the streets, congregate with their friends, and use the town as their own playground. Because Sparrows Point was small and tightly controlled by the company police, there were few dangers threatening children. Steelworkers' coupon books allowed children to use the company store as a place to shop without their mothers' supervision. Crystal Young remembered the freedom and fun times: "I would treat all the kids to ice cream and charge it to my father's coupon. All of us did it."

The company store also meant that Sparrows Point wives could turn some of the responsibilities of shopping over to their children. In a small, self-contained community where people knew one another well, the casualness with which the company store operated meant that mothers could let their children do their clothes shopping without adult supervision:

> Even at a young age, say twelve, you'd say, 'I'm going over to the company store,' and our mother would say, 'O.K.' You'd go over there and they would have gotten new dresses in and I'd pick out three or four dresses and I'd say to the lady, 'Could I take these home on approval for my mother?' She'd say, 'yes,' and they wrapped them, no charges, and I'd come home and show them to my mother and she'd say, 'O.K., you can keep them.' I'd run back and tell the clerk.

Sparrows Point wives were relieved of considerable worries because the town was self-contained, everything was within easy traveling distance, and children could do errands or negotiate the streets without fear of danger.[25]

FIG. 5 Sparrows Point managers pose with the wives and daughters of steelworkers, c. 1900.

The standards of behavior in Sparrows Point were established by the steel company and complied with by residents who both needed their jobs but also valued the atmosphere of harmony that existed. Since the town of Sparrows Point was owned and controlled by the steel company—first Maryland Steel and eventually Bethlehem Steel—the company established the rules and had its own police to enforce them. The people who remained in Sparrows Point, particularly those families who remained for a second and third generation, appreciated the conformity imposed by the company because there was always the feeling that things were safe in Sparrows Point: "Nobody ever said, 'Be careful' down there because you were on your own. I never in my whole life down there felt that it wasn't safe. One time one policeman fired a gun. Someone was trying to break into the milk truck. But there was never, never any violence on Sparrows Point. None of us ever had any doubts." The company town residents who never "had any doubts" were most likely to be those ensconced in a stable and settled family.[26]

Sparrows Point offered different amenities for different groups of residents. For married men who were raising families, the company town offered the stability of work that was steady, within walking distance from home, and that paid wages sufficient to live comfortably in the housing maintained by the company. Single men who were boarded or lived in the shanties often endured long periods of loneliness in order to send their wages back to hometowns. For African Americans there were houses, schools, and churches, all provided by the company, but most of all there were industrial jobs during an era when black men were systematically excluded from most of America's industries. The company provided affordable housing, a store where workers could charge items against their wages, and a police force that maintained order. The strict enforcement of the company's rules are generally remembered as beneficial by people who had carefree childhoods in Sparrows Point and who themselves became steelworkers or the wives of steelworkers.

However, the principles of paternalism that characterized Sparrows Point, as well as other company towns of the late nineteenth and early twentieth centuries, had their restrictive, even repressive, effects. If a steelworker were fired for coming into conflict with the mill's management, his family would lose not only their source of income but also their home. For union organizers, the town of Sparrows Point was impenetrable. Owned and maintained first by the Pennsylvania Steel Company, and then from 1891 to 1916 by the Maryland Steel Company, and finally bought in 1916 by the Bethlehem Steel Company, both the steel facility and the town were tightly controlled by the owners.[27]

The voices of those who were evicted from the Point are difficult to retrieve. Certainly injustices must have occurred and women and children must have suffered, at times, for the unacceptable habits and behaviors of a father or older brother who drank or fought, or was caught working on union activities:

> If they found out that anybody was in trouble, if you were some kind of a delinquent or something, your family had to move off the Point. You had to. They kept you in line. Everybody knew everybody; therefore you were afraid to be bad. You didn't do anything really out of line. So the police didn't have that much to do. One time one policeman caught a man going to work drunk. The punishment? Well there wasn't really any. The policeman took the man home to sober up and he missed a day of work. That's the worst thing I remember ever happening.

Former residents of the company town relished the safety and security of their now-demolished community, but they also acknowledged how pervasively the company maintained control and conformity.[28]

After 1900, increasing numbers of steelworkers commuted to the Point. A streetcar line built in 1903 gave immigrants from East Baltimore access to jobs that had been the monopoly of the native-born whites and African Americans who had settled in the company town.[29]

During World War I two steelworker communities, Dundalk and Turner Station, were built on the peninsula directly west of Sparrows Point. Dundalk began as housing for white workers at the Sparrows Point Shipyard, and the plats show a configuration of streets with names like Bayship, Portship, and Kinship, laid out in curved avenues to form the bow and stern of a large tilting ship. In 1917 Bethlehem Steel commissioned the Dundalk Company to continue the construction of homes in Dundalk that could be bought by white steelworkers through a payroll deduction plan. Five hundred homes of gray stucco with slate roofs were located on tree-lined streets, and in the center of Dundalk an attractive shopping center was built around a grassy park. Subsidized housing was also built under government auspices for war workers in Turner Station, but Bethlehem Steel chose not to invest in an architecturally coherent community for African American steelworkers and their families who found it necessary to be much more self-sufficient in the development of their community.[30]

Turner Station, a community built on rural, waterfront property on the southern end of the Patapsco Neck, had only ten residents in 1918. The subdivisions of Carnegie and Steelton Park were platted out in 1919 in an area southeast of Dundalk, and individual African Americans who worked at the Point built private homes in Turner Station, financing them either with their own savings or with mortgage loans from black self-help associations. Local black-owned businesses were built, including restaurants, grocery stores, barbershops, and a movie theater, along with churches and schools. Eventually, Turner Station became one of the largest black communities in Baltimore County.[31]

Ultimately, the histories of the two steelworker communities would include notable similarities, because Dundalk and Turner Station both grew in the shadow of the steel industry that dominated the area. They developed into middle-class communities based on a strong work ethic and the wages men earned from industrial jobs at Sparrows Point. Both communities established traditions of homeownership, economic security, vigorous religious and educational institutions, and award-winning recreation leagues.[32]

The Depression brought to an end the economic growth connected with World War I. Between 1929 and 1932 employment at the Point dropped from 18,000 to 3,500. Even those men who kept their jobs at the Point were working only two or three days a week. Wives of steelworkers who lived in the town of Sparrows Point were able to fix modest meals for their families because Bethlehem Steel gave those families credit at the company store along with extensions on their rent. At the same time, however, the Bethlehem Steel Company took a stand against a proposed state-sponsored system of unemployment insurance. Working steelworkers helped in private charitable drives, including a call for food, clothing, fuel, and shoes from Post 88 of the American Legion, Sparrows Point, to help some of the 100,000 people reported in the local press to be almost destitute.[33]

The hardships of the Depression helped spur the drive to organize steelworkers. The Steel Workers Organizing Committee (SWOC) had been successful in getting Big Steel companies such as U.S. Steel and Jones and Laughlin to sign contracts with the United Steelworkers of America (USWA) in 1937, but that same year Bethlehem Steel vowed to fight unionization and hired 300 special police armed with guns and tear gas to quell labor agitation in Baltimore. At the Sparrows Point plant, the company town was a source of resistance to the union. Steelworkers living in Sparrows Point were reluctant to participate in union activities because they were vulnerable to losing the considerable amenities of affordable housing close to both the mill and the company store, as well as the crucial economic benefit of credit on food and rent during hard times.[34]

Union organizing among Bethlehem Steel's production workers in the 1930s had its headquarters in the Finnish Hall in the Highlandtown section of Baltimore. Finnish, Polish, and Italian immigrants from the Highlandtown section of Baltimore and African Americans from Turner Station and West Baltimore carried the burden of the most difficult union organizing, because Bethlehem Steel banned union activities from the company town of Sparrows Point. Dundalk was also relatively inaccessible to union organizers because it had close ties to the steel company, and in the 1930s it was the preserve of higher-paid skilled workers and foremen, men who tended to oppose the union.[35]

Union activism was centered in communities adjacent to Sparrows Point, including Turner Station, Essex, and Edgemere in Baltimore County. Highlandtown, a neighborhood of working-class immigrants in East Baltimore and the home to many steelworkers, became the center for SWOC organizing. Local organizers determined that because one-third of the paid workforce at the Point was African American, it was essential to unite black and white steelworkers in a joint effort to improve their wages and working conditions. West Baltimore, home to many African American steelworkers, was another center for union organizing and support. The union campaign also provided leadership positions for the wives of steelworkers, who were active in many aspects of the campaign.[36]

Bethlehem Steel fiercely resisted unionization until World War II, when arms manufacture brought production and employment back up to levels even higher than in 1920. Rising profits and a fear that strikes would interfere with production weakened the resolve to keep unions out. In 1941, the USWA won their unionization drive at Bethlehem Steel's Sparrows Point plant by a vote of two to one after a long and difficult union drive. The collaboration of white and African American steelworkers on the campaign to win USWA representation is a notable example of blacks and whites working together at a time when most institutions in America, even the U.S. Army, were segregated.[37]

Dundalk tripled its population during World War II, when large numbers of migrants came to Sparrows Point for jobs. White workers, referred to by Dundalk natives as "farm people and small town people," came from West Virginia, Virginia, Tennessee, New Jersey, and Pennsylvania, expanding what had been a close-knit "village" into a larger and more heterogeneous town.[38]

Turner Station accommodated African American war workers with six new housing complexes. One of these projects, Anthony Homes, was referred to as the first "colored and privately owned housing project in Maryland." Anthony Homes was financed by the local African American physician, Dr. Joseph Thomas, and managed by his wife, Flavia Thomas. Carver Manor Homes was financed by the state of Maryland as rental property and later sold to individual homeowners. Day Village was built in 1944 and is still owned by the son of the original developer. An advertisement published in 1955 recruited homebuyers by describing Turner Station as a "self-contained Negro community" with a "very low rate of Juvenile delinquency."[39]

For eighty years Turner Station and Dundalk sat side by side, each growing in population and each developing completely separate churches, social clubs, recreation leagues, restaurants, bars, and shopping districts. The two communities shared aspirations for individual home ownership and economic advancement. The history that links the larger Dundalk community to Turner Station is one of hardworking women and men engaged in com-

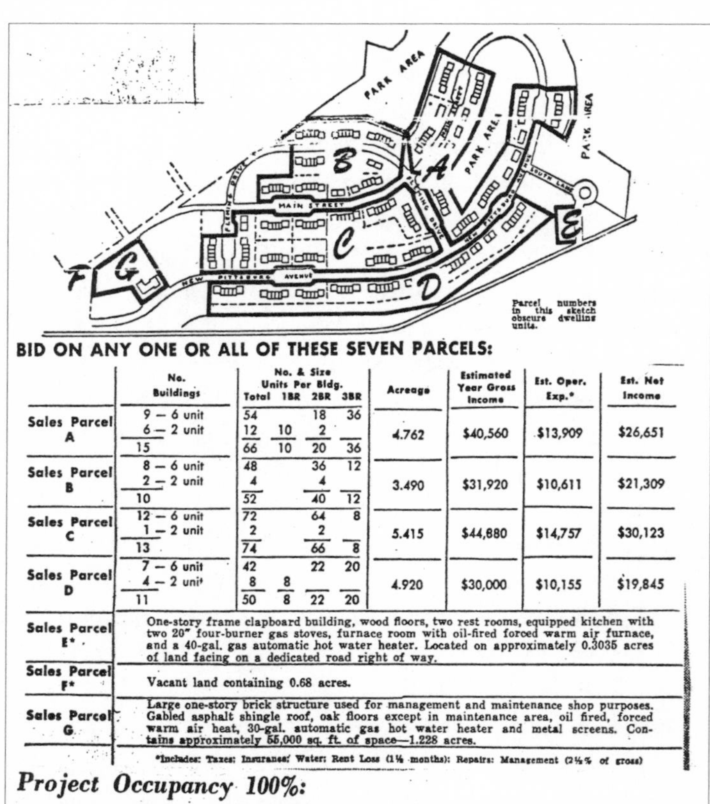

BID ON ANY ONE OR ALL OF THESE SEVEN PARCELS:

	No. Buildings	No. & Size Units Per Bldg. Total	1BR	2BR	3BR	Acreage	Estimated Year Gross Income	Est. Oper. Exp.*	Est. Net Income
Sales Parcel A	9 — 6 unit	54		18	36	4.762	$40,560	$13,909	$26,651
	6 — 2 unit	12	10	2					
	15	66	10	20	36				
Sales Parcel B	8 — 6 unit	48		36	12	3.490	$31,920	$10,611	$21,309
	2 — 2 unit	4		4					
	10	52		40	12				
Sales Parcel C	12 — 6 unit	72		64	8	5.415	$44,880	$14,757	$30,123
	1 — 2 unit	2		2					
	13	74		66	8				
Sales Parcel D	7 — 6 unit	42		22	20	4.920	$30,000	$10,155	$19,845
	4 — 2 unit	8	8						
	11	50	8	22	20				
Sales Parcel E*	One-story frame clapboard building, wood floors, two rest rooms, equipped kitchen with two 20" four-burner gas stoves, furnace room with oil-fired forced warm air furnace, and a 40-gal. gas automatic hot water heater. Located on approximately 0.3035 acres of land facing on a dedicated road right of way.								
Sales Parcel F*	Vacant land containing 0.68 acres.								
Sales Parcel G	Large one-story brick structure used for management and maintenance shop purposes. Gabled asphalt shingle roof, oak floors except in maintenance area, oil fired, forced warm air heat, 30-gal. automatic gas hot water heater and metal screens. Contains approximately 55,000 sq. ft. of space—1.228 acres.								

*Includes: Taxes; Insurance; Water; Rent Loss (1½ months); Repairs; Management (2½% of gross)

Project Occupancy 100%:

Lyons Homes has been constantly occupied since construction at the 100% level, with a waiting list at all times. Average rent is $49.17. Rental delinquency has been practically non-existent. 52% of the families have been in occupancy since 1942. Rate of tenant turnover does not exceed 4%. Vacancy since construction has been less than 1% aggregate average.

Bidding Specifications:

Bids may be submitted separately on Parcels "A", "B", "C" and "D", or on all four parcels collectively, or on any combination of the four. If you submit a bid on a group of parcels, but also wish to be considered for any part of the group, you must specify individual parcel prices. Parcels E, F, and G, must be bid on individually. *Terms are offered on Parcels E and F only if sold with another parcel or parcels. Sealed bids will be opened at 2:00 P.M., April 25, 1955, in the *Washington office listed below.* For bid forms, inspection of properties, or further information you may call, write or visit MR. ROBERT FERGUSON, PROJECT MGR., ATWATER 4-0900, 411 NEW PITTSBURGH AVE., at the Project. Or you may call or write the Washington office listed below.

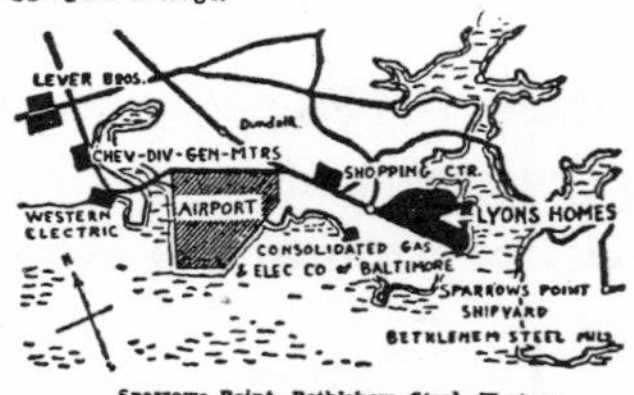

Sparrows Point, Bethlehem Steel, Western Electric, Glidden Paint Co., Union Oil Co., Chevrolet Baltimore Division, American Standard Plumbing, and Lever Bros., are some of the plants in the vicinity of Lyons Homes.

FIG. 6 Turner Station had considerable rental property, such as Lyons Homes, built to accommodate the influx of African American workers during World War II.

mon work experiences. However, the common experiences shared by Turner Station and Dundalk have, until very recently, always been across segregated social boundaries.

Today, Turner Station and Dundalk are communities that are struggling with the economic changes associated with deindustrialization. According to the 2000 census figures, the 75,500 mostly white Dundalk residents constitute an eleven to one numerical majority over the smaller community of Turner Station, which has 3,000 mostly African American residents. This is a larger white majority than was characteristic of the original demographics of the community. According to the 1900 census, Sparrows Point included 2,404 white residents and 1,053 African American residents, which meant that African Americans composed fully one-third of the company town of Sparrows Point.[40]

The loss of steel jobs affected Turner Station in some ways that are distinctly different from the impact of deindustrialization on Dundalk. The thriving African American community of Turner Station is much smaller than Dundalk, both in geographic size and in population. With nowhere to expand and with housing in Dundalk unavailable to blacks until the 1990s, the middle-class children of Turner Station steelworkers were more likely to move to the African American neighborhoods in Northwest Baltimore. Communities like Randallstown provided the suburban housing desired by educated black professionals who were the children of Turner Station's steelworker families.

In other ways, however, the impact of deindustrialization on Dundalk and Turner Station has been similar. A primary consequence of the postindustrial economy of southeast Baltimore County has been families reverting to households with more than one source of income. In steelworker communities that used to support breadwinner/homemaker families, there are now two, and often three or more, paychecks supporting a household where adult children or retired parents may be contributing to the household budget.

The single most significant change is that wives have gone into the full-time paid workforce in large numbers. The wives of steelworkers who in 1900 managed boardinghouses for single men in Sparrows Point today seem closely akin to contemporary wives who are managing day care centers, bank branches, business offices, and hospital units. Deindustrialization has also played a central role in transforming the gender relationships that were characteristic of the 1950s and 1960s, when the preponderance of the economic power rested with a breadwinner husband.

Today the communities surrounding the Sparrows Point steel mill are home to a wide variety of households. For some steelworker families the loss of jobs in steel can be directly connected to divorce, domestic abuse, or illness. For younger generations of women and men, job prospects require that they get specialized training or higher education, without which they face the possibility of being stuck in low-paying jobs. *Wives of Steel* focuses, however, on those households where women have gone into the paid workforce in order to compensate for the loss of high-paying jobs in steel that in the 1950s and 1960s supported breadwinner/homemaker families.

2

The Gendered World of Steel: It's a Man's World Inside the Sparrows Point Mill

When I was working at the Point I called it "The Rock," because it was hot, dangerous, dirty work, and the hours were long. A lot of the guys were boisterous and loud, because there was so much noise down there that they learned to talk loud to each other. They were competitive with each other. The steel side, the finishing side, the slaggers in the open hearth, the riggers, would taunt each other about, "Oh man, you guys don't do anything over there, just girl stuff. Try coming over to where we're working if you want to see some real work."

—JOHN SPRIGGS, JULY 1989

To understand the lives of steelworkers' wives in Baltimore prior to World War I, it is necessary to understand the work environment that awaited the Sparrows Point steelworkers. Not only were most of the jobs at the Point grueling, it was also a place where several thousand men worked together without the intrusion of women. At the Point a code of behavior developed that emphasized the manly qualities required to work in steel. Under these conditions, sharply gendered social worlds resulted, and at the ends of their shifts steelworkers brought home the stresses of what was considered the most macho of American industries.[1]

The ethic inside steel plants, as was true aboard ships and inside mines, was that women must not be allowed to enter the mill, based on the superstition that their mere presence could cause an accident or a fatality. The paid workforce at Sparrows Point, already exceeding 2,000 men just thirteen years after the mill was designed, worked together amidst harsh, unhealthy, often dangerous conditions, contributing to a masculine environment of both camaraderie and competitiveness. This masculine environment was reflected in quite different ways in the mill and in the company town.[2]

At the turn of the twentieth century, married and single steelworkers were positioned differently in the company town of Sparrows Point. For men who were married, manliness was based in part on their roles as husbands, fathers, and heads of households. The role of provider was a significant one for family men who accepted the hardships of working in a steel mill in exchange for the wages, comfortable housing, safety, and conveniences that Sparrows Point afforded their wives and children. But for single men—the majority of the town—there were problems of loneliness, and particularly for men living in the shanties, the poor living conditions often led to drinking and fighting.

For single steelworkers living in the abysmally cramped shanties, recreational rowdiness came to characterize the little bit of leisure time that was permitted after the long shifts at the Point. Rufus Wood had planned for a sober, industrious workforce, but in April 1890 he wrote to his brother Frederick complaining about outbursts of rowdiness that plagued his company town: "I drove by Dorsey's yesterday and they were having a regular circus & carusal [*sic*]. Men lying in the gutter & on the fences, others yelling & hooting & Birmingham complains of the noise and disorder, while the neighbors are afraid to go by especially after dark. . . . I[I]t appears that a good many cases of bad women especially across the creek have just turned up." Dorsey's was one of several taverns located beyond the property owned by the Maryland Steel Company, which outlawed alcohol on its premises. Dorsey's was a place where steelworkers from the Point consumed alcohol in defiance of management's regulations. Furthermore, "across the creek" was, of course, where hundreds of single men, native-born white men, immigrants, and African Americans, were warehoused in shanties. Without families, without access to the recreational accommodations on the other side of the creek, some of these men sought pleasure in women and drinking.[3]

Drinking and carousing with "bad women" were two of the ways in which the most marginalized single steelworkers participated in masculine rituals that proved one's virility. In a company town that included nearly fifteen hundred single men working long hours under difficult conditions, isolated from the recreations of an urban area, the existence of prostitution within three years of the founding of Maryland Steel should come as no surprise. In his role as industrial planner, Rufus Wood had expectations for workers' behavior that were unrealistic given the demographics of a town where men outnumbered women three to one.[4]

In June 1889, less than two years after beginning the construction of housing on Sparrows Point, Rufus warned his brother of the need for a jail in the company town: "We have immediate need of a strong lock-up where unruly prisoners can be confined until disposed of by a justice of the peace. The only alternative being a long ride to Highlandtown twenty miles up and back with each offender, and absence of the officers meantime." Rufus blamed the problems on excessive drinking and a laxness in the policy he himself wanted enforced of firing men who made a habit of getting intoxicated: "It is getting to be a well-understood fact of late that men will not be discharged for drunkenness alone. And they make less & less attempt to conceal their condition. We had no such trouble the first year, because men understood that repeated drunkenness involved discharge." In Wood's letter we get the first reference to a workplace dynamic that would characterize the Point throughout its history. Individual mills and departments became the province of different ethnic, racial, or religious groups, in a hierarchical arrangement that privileged certain groups while at the same time other groups were excluded or denied access to better jobs.[5]

Rufus took it for granted that white men would be favored over black men, and that native-born men would be favored over immigrants from Eastern Europe, but in 1889 what he complained about was religious rather than racial or ethnic discrimination. From his Protestant frame of reference, Catholics were the recipients of a pattern of favoritism that threatened the wholesomeness of the working and living environment at Sparrows Point:

> [T]he foremen easily talk Col. Franklin over into retaining men for whom they have a particular fancy. Many of whom are the worst characters in the place. I can name twenty men you would not allow to be in the employ of the company at Steelton because of their confirmed habits of intoxication and who are making this place anything but desirable [as] . . . a home for families with women & children. And in confidence I will tell you what I have not even told the Colonel that in the discharge of men, Catholics

> are always favored as against Protestants in the gangs and the proportion of that sect is largely increasing.

Rufus's letter failed to result in a satisfactory solution, and favoritism and discrimination continued to plague the Point through the twentieth century, as did problems with drinking and verbal and physical rowdyism.[6] In a community isolated from other communities and overwhelmingly composed of young males, almost three-quarters of them single, an ethos of manliness emerged that centered on the qualities of strength and endurance; the willingness to face the high risk of industrial accidents; the temptations of rowdiness for men living detached from family structures; and, for men with families, the role of provider.[7]

The Bessemer process of producing steel dictated that furnaces heated to between 2900 and 3000 degrees Fahrenheit cannot be shut down at the end of the day or at the beginning of a weekend. Like other steel mills, the Sparrows Point steel complex has always required an industrial process that continues twenty-four hours a day, necessitating a large workforce on shifts. When operations began at Sparrows Point in the late 1880s, the labor force was organized into only two shifts scheduled according to a system designed to maintain operation of the mill twenty-four hours a day, seven days a week. The day shift worked eleven hours a day, seven days a week, while the night shift worked thirteen or fourteen hours a day, seven days a week. The swing shift occurred on Sundays when the night shift worked twenty-four hours straight through, thereby "swinging" over to function as the day shift. There were no vacations and the only two holidays, Christmas and the Fourth of July, were unpaid.[8]

The long turns of between ten and fourteen hours and the twenty-four-hour swing shift were partially alleviated in the 1920s, when President Warren Harding and the American public pressured the large steel companies to reduce the horrendous hours of steelworkers for humanitarian reasons. In 1923 the major steel companies, including Bethlehem, grudgingly agreed to eliminate the twenty-four-hour turn and to institute the eight-hour day and a six-day week. They qualified the agreement, however, by saying, "as soon as the labor supply permitted." This allowed mill managers to claim that a shortage of labor necessitated squeezing the greatest number of hours out of their workers.[9]

The twenty-four-hour swing shift ended, but there were still areas of the mill outside of the open hearth and steel rolling departments where men continued to work ten or more hours each day. Consequently, weekly hours in American steel mills were cut back from 84 to between 56 and 70, and as late as 1929 a quarter of the steelworkers at Sparrows Point were still working seven days a week. Although much has changed in steel manufacturing, including technology, safety measures, and ultimately women working in production, the masculine identity associated with working at the Point continued to be based in part on the long hours and grinding working conditions.[10]

Shift work has always been a part of steelmaking because of the necessity of running the production process continually, but it is an aspect of this industry that has most interfered with steelworkers having time to build satisfying relationships with the members of their families. Rodney Anderson, who worked in the plate mill for twenty-two years, described his own regularly rotating schedule:

> I'm lucky to have a regularly rotating schedule, because I have a four-day weekend off once a month. But even with regularly rotating shifts, when I rotate from the 3 P.M. to 11 P.M. shift to the 7 A.M. to 3 P.M. shift, it just ruins my sleep. This is what my schedule looks like:

SUN	M	T	W	TH	F	SAT
OFF	OFF	3–11	3–11	3–11	3–11	3–11
3–11	3–11	OFF	OFF	7–3	7–3	7–3
7–3	7–3	7–3	7–3	OFF	11–7	11–7
11–7	11–7	11–7	11–7	11–7	OFF	OFF

This work schedule has some advantages, but also major disadvantages, for steelworkers and their wives. In the course of one month this steelworker works seven days in a row from 3 P.M. until 11 P.M. Then he has two days off and returns to work seven days in a row from 7 A.M. to 3 P.M. Finally, after one day off he works seven days in a row from 11 P.M. to 7 A.M. The reward comes with a four-day reprieve from work. Over the course of the month, however, this schedule requires a steelworker to change work and sleep cycles three times. His wife must arrange meals to accommodate these changes, keep children quiet during the days when he works the eleven-to-seven shifts, and absorb the stress that accompanies the sleep deprivation.[11]

Other work schedules can be more or less routine, depending on the type of job, the part of the mill in which a steelworker is assigned, as well as whether or not there is a break in production that requires repairs. Mark Kramer, who is in his early 40s and who has worked as a laborer, a machinist, and now an electrician at the Point, feels fortunate to have a straight daylight job after 26 years of working many different kinds of schedules in the mill:

> Some steelworkers do work a regular daylight shift from 7 A.M. to 3 P.M. They are mostly maintenance workers and lower-paid laborers. A relatively small number of workers are on permanent 3 P.M. to 11 P.M. or 11 P.M. to 7 A.M. shifts or "turns." The rest of the plant population works rotating shifts to cover continuous production lines. Some only learn their schedule on Thursday for the following week. They could be scheduled any day, any hour.

The absence of a routine schedule makes the life of a steelworker unpredictable and can cause digestion and sleep disorders. A swing shift has particular meaning in this community and is distinguished from working straight daylight or working straight night shift, because the swing shift is a work schedule that swings between day and night and is most disruptive to sleep and family life. Mark explained how his own schedules had changed in different areas of the mill: "I was a laborer and there were times when I would actually work all three shifts in one week. The company wanted 24/7 coverage, whatever the job was. Later I went into a machinist apprenticeship and I did a month of daylight, a month of 3 to 11, and a month of 11 to 7. Still never got used to it. I recall some days that I didn't know what day it was." Mark expressed with resignation how it feels working the 11 P.M. to 7 A.M. shift: "Even if you stay in bed for 12 hours the next day, you wake up feeling like you didn't have any sleep. Anyway you cut it, with shift work you don't get enough sleep."[12]

Steelworkers are frequently asked to work overtime, and many are eager to earn overtime pay of time-and-a-half. The consequence is a high degree of unpredictability in work schedule, and almost no control over determining which days he gets off. Also, steelworkers continue the practice of working double shifts if a replacement doesn't show up or if there is an emergency or a particularly urgent deadline pending. A double means a total of sixteen hours straight on the job.[13]

Swing shifts and being asked, usually at the last minute, to work a double shift are the most unpredictable parts of a schedule for steelworkers. For steelworkers in the production area they are also the best opportunities to earn overtime. The consequence for the wife of a steelworker is the necessity of being at home during all of those periods of time when children, parents, in-laws or even pets need attention, because she will never know for sure when to expect her husband to be at home. The swing shift system makes it impossible for a family to plan on a stable weekly schedule throughout the year, and it makes the wife essential to maintaining the home life, meals, and child care that keep the family intact.[14]

The dangers of employment at the Point are apparent, and everyone who works there knows of a steelworker who has been electrocuted, swallowed in a tank of acid, or caught in a piece of machinery. Jim Baneck, a cold roller who worked for twenty-four years at the Point, described a particularly grisly accident in the number 56 pickler, a part of the plant where deadly chemicals are used to oxidize the surface of steel during the finishing process: "This guy was standing on lumber that covered a tank filled with sulfuric acid. The tank cover was being changed plank by plank. When he stepped on one of the old planks it gave way and he fell into the acid and was killed. Everybody was shocked by it. We all were thinking, 'My God, that could have happened to any one of us.'" Steelworkers who know their labor history are quick to point out that during World War I more Americans died in industrial accidents than the 116,000 who died on the battlefields of Europe.[15]

Safety issues add another dimension to gender roles at the Point. Peggy DePaulo, who worked in production for thirty years, was one of a small number of women in the mill at the Point, yet she was the one to assume some leadership for improving safety. She observed that in the 1980s there was some reluctance on the part of men to voice individual complaints about safety:

> When I agitated for better safety measures, the men I worked with were appreciative. But they aren't about to cry for help whenever there is a dangerous situation because that's a sissy thing. Maybe it says, "I'm not man enough to work in these conditions." At the Point there is a hesitation on the part of the individual men to complain about something being dangerous.

The ethos of manliness at the Point made it more acceptable for a woman to advocate for safer work conditions, because it might appear unmanly for male steelworkers to complain about dangers that needed remediation.[16]

There are also many parts of the Sparrows Point mill where steelworkers are exposed to industrial chemicals and dust, and to intensities of heat and cold. A man working as a mechanical engineer in the centering room casually described the pollutants he is constantly exposed to: "It was hot and dirty, and everything and everybody was covered with steel dust. At the end of every day I was covered with steel dust. You took a shower before leaving the place because you were filthy with sweat and steel dust by the end of your shift." The long-term effects of this environment are high rates of cancer and asbestosis.[17]

Jean Edwards spoke for many steelworkers' wives when she recalled the job of washing her husband's work clothes: "I remember the clothes. I remember the smell in the clothes, and I remember the just downright dirtiness. I have never seen such dirty clothes. It was just really, really hard work, hard, sweaty, dirty work." Mildred Frantz grew up observing her father and mother go through the end-of-shift routine: "My father came home at the end of his shift covered with red dust and dead tired. My mother made him peel off his clothes at the door so that he wouldn't get the whole house dirty. Then he would take a bath and fall into bed like a dead man. That's what I remember of my father." Sarah Parker saw the grime from her husband's work at the Point in the bottom of her washing machine: "It needed all sorts of scrubbing to get his clothes near to clean. I would rather have taken it to the laundromat because it was so full of dirt and soot that after one load of his work clothes and my machine would be covered with grease."

While a steelworker's wife is focused on the grime that she has to launder, many steelworkers are wearing their "uniforms" proudly. Hardhats are a safety requirement and casual clothes make sense in a workplace filled with steel dust, but Eric Logan, a veteran of the pipe mill, emphasized the image created by the clothes that steelworkers wear: "At the Point you could dress the part—dungarees, steel-toed boots, plastic helmet, cut-off shirts. It gave the appearance of a real macho." He went on to reflect that for many men at the Point working in the steel industry is confirmation of robust manliness:

> A lot of guys who work at the Point are still in the stage of improving their manhood. They can say, "Hey, I work in a steel mill." What better image of manhood is there? It's dangerous and its hard physical work. You're not going to have soft hands. Why, just the word "steel" implies toughness. Look at Superman; he's the "Man of Steel."

When steelworkers talk about the Point, their description of their work is peppered with references to the extremes found inside a steel mill, and complaints about the hardships of steelmaking are interwoven with boasts about their own prowess and toughness. Indeed, when steelworkers from the Point talk about their work, the sturdiness of the worker and the toughness of the steel they manufacture blend together in an expression of pride based on both the skills steelworkers demonstrate and the "macho" reputation that steelworkers are granted by the larger world.[18]

There is camaraderie and combativeness among steelworkers at the Point, based on bonds of common experiences as well as antagonistic divisions. Crews of three or four guys who worked together regularly have often been close to one another, giving each other rides to work, socializing together after a shift, and most importantly looking out for each other in dangerous situations. The closeness of a work crew could also become an obstacle to a steelworker's involvement with his wife and children since, as a thirty-year veteran of several departments said, "Sometimes you spent more time with your work crew than you did with your own family."[19]

The Point is a workplace where the rough and the boisterous set the standard for a work culture where hazards and hardships go hand-in-hand with a playful rowdiness that is expressed in pranks, practical jokes, and name-calling. The discourse within the mill has been characteristic of a masculine work culture that suggested, "this is men's work, can you handle it?" Howard Cook explained that men who are new to the Point are put through an initiation that tests their suitability to the toughness in the mill: "When a new employee comes on to one of the mill crews, we look them over and give them a hard time. Somebody will say something like, 'Look what the cat dragged in today,' letting them know that they'll have to prove themselves to fit in." Howard saw a system that "baptizes" workers at the Point as members of a family of steelworkers, nicknames are assigned that make each worker stand out as a character in the ongoing drama of making steel:

> Everyone gets a nickname in the steel mill, because over the years a lot of characters have gone through there. The superintendent had a nickname, but no one called him by his nickname to his face. His nickname was "Pinky" because he had a red mustache. Then there was "Glow Worm," "Road Apple Andy," "Bones," "Locker Room Lou," "Powder Room Pete," "Rubber Jaws," "Squirrel," "Twitchy," "Hot Rod," "'Tater Head," "The Preacher," "Green Bean," and "Rag Top." My nickname was "The Hatchet Man."

In these nicknames lie clues to affection as well as antagonism. Nicknames are a way of making work crew relationships more familiar and congenial by identifying distinctive characteristics that elevate a man out of the category of generic factory worker to the status of a person with unique and therefore special qualities. With subtle ambiguity steelworkers can also use a nickname to rein in undesirable behaviors in a coworker. Rubber Jaws talks too much, and he can continue to do so, but he will be reminded on a daily basis of his excessive talking.

Nicknames also help steelworkers cope with the rigidly hierarchical structure of their workplace. Having a nickname for the superintendent, no matter how benign, is a way of cutting him down to size. Calling the superintendent "Pinky" is a way of giving him a diminutive "handle" or label. The name may literally refer only to his red moustache, but it implies a lot more. After all, the pinky is the smallest finger on the hand. The nickname "Pinky" suggests feminine references that diminish the superintendent's titular status, as opposed to "Red," which as a nickname would render more social power. Keeping the superintendent's nickname secret is a way in which steelworkers control a piece of information that is unknown to management.[20]

The language of manliness at the Point also includes the use of profanity. Mildred Frantz thought of it as a work environment where most things were rough and where historically there were no "ladies" to defer to:

> There was a mentality in Bethlehem Steel. Guys had the foulest mouths. They cursed like sailors. Sometimes I would go places and I would hear people talking and I would end up asking them, "Do you work down at Sparrows Point?" and they would be surprised and say, "Yeah, do I know you or something?" No, but I just knew where they worked from the conversation. There is a mentality that went across the whole place.

There have been plenty of steelworkers at the Point who did not curse, and even some senior workers or lay preachers who could command enough respect to suppress foul language among their crews, but cursing was by and large part of the work culture.[21]

Camaraderie among work groups contrasts sharply with combative divisions between groups, which could be based on several factors, including whether you work on the steel side or the finishing side, or whether you are a white hat or an hourly worker. It could also be based on ethnicity, race, gender, position in the plant hierarchy, or simply on personality.[22]

Accounts of the role of alcohol in the lives of steelworkers at Sparrows Point recur frequently, beginning with Rufus Wood's complaints in 1888—just one year after the mill opened—that workmen were leaving the Point to drink, and that certain men were involved in chronic incidents of intoxication. The use of alcohol among steelworkers has a history similar to that of other industrial workers. The grinding work schedule resulted in exhaustion after the long shifts and gave the traditional stop at the local bar an important role in the lives of many steelworkers trying to relax and unwind.

Men and women who grew up in Sparrows Point remember that it was a common practice to flaunt the company's rules against the sale of alcohol within the limits of Sparrows Point. Vince Cipriani remembered an array of taverns, bars, and nightclubs were located right beyond the limits of the company town and were within easy walking distance of most homes on the Point:

> The First and Last Saloon in Edgemere had that name because it was literally the last place where men could buy alcohol before coming onto company property, and the first place where they could buy alcohol when leaving the Point after a shift. Another place was The Green Castle, a restaurant and bar on North Point Road. Mickey's was right near by and is there to this day. Alcohol wasn't allowed to be sold anywhere in Sparrows Point, but that didn't keep men from leaving the town and going to one of the places close by.

Drinking, if it was not excessive and did not lead to public disturbances, was a violation of company rules that aroused little disapproval for most people living in the company town.[23]

Whether he is a drinker or not, every man who has worked at the Point is familiar with the bars that sit immediately outside the gate. In the tradition of Dorsey's tavern, bars like Mickey's, Gail's, Hogameyer's, Pop's Tavern, The Jolly Post Inn, and The White House functioned to lure steelworkers from the Point to drink and relax at the end of a shift. Until the stringent drunk driving laws of the 1980s, drinking was as much a part of the work environment of a steelworker as steel-toed shoes. Jim Baneck, who started in the machine shop in 1963, described the obligatory stop at a local watering hole after the shift: "Drinking is a part of the mystique of being a steelworker. Every guy feels overworked, and this is a good way to relax as well as a chance to indulge in a lot of bravado—joking and challenging and bragging about sexual prowess. I can remember standing on the lot at Mickey's challenging some guy to drink a pint faster than me." Steelworkers report stopping at bars at the end their shifts, regardless of the time of day. Bars opening at 6 a.m. allowed even crews from the night shift to stop for their ritual drinking session before going home. No man was excused from the obligation to socialize with his crew at the end of the shift, and frequently that meant drinking: "If you were a teetotaler, you couldn't make it at the Point. You would have to isolate yourself. Drinking together is that much a part of being a steelworker."[24]

Some steelworkers report suffering from alcohol-related problems that they associate with their work at the Point. Scotty Fraser, a veteran roller, explained the pressure to abuse alcohol by recounting the story of one of his coworkers: "When this guy started to work on my crew he didn't drink at all. But the first day, after the shift, the other guys were just begging him to go to Gail's for a drink, and when he ordered Coke they laughed and hooted at him until he finally caved in and drank a whiskey. That guy is a stoned alcoholic today; I watched it happen." Interviews with the adult children of steelworkers from the Point include many accounts of devoted family men with little interest in drinking, and many retirees reported that their crewmates' rituals did not include alcohol, but for the thousands of men who have worked at the Point, alcohol has been no stranger.[25]

Combativeness at the Point also found expression in a racial tension and antagonism that historically has divided black and white workers in the competition for higher-paying, more desirable jobs, and an ongoing effort to maintain a pecking order that has an acceptable place for white men.

African American workers have experienced a long history of being excluded from higher-paying, skilled jobs at the Point. In his initial design of the mill, Rufus Wood identified black men as the workforce of choice for the hottest, most strenuous laboring jobs at Sparrows Point. And because of Jim Crow practices that persisted until the 1970s, white steelworkers were able to hold tenaciously to their monopoly of the best jobs at the Point.[26]

For most of the mill's history, only a handful of blacks worked in any job category other than laborer, and John Spriggs described the coke ovens—the hottest and dirtiest area of the mill—as all-black departments that came to represent a pariah status for African American steelworkers: "They put us in the coke ovens, which is like working in hell. There is constant smoke, constant gas, constant pollution, and the smells are sickening. You are always covered with grime, and from the day you start there until the day you quit, it never gets out of your skin." African American steelworkers cite numerous examples of having been bypassed for a promotion that went to a white man. The universal lament is, "Why did the white guy come in eighteen and clean-faced and get all of the good jobs?"[27]

A system of white cronyism prevailed until 1974. Ron Moore recalled the widespread belief that skilled jobs at the Point were most appropriately an employment resource for white men living in the surrounding communities: "White guys who went to the local high school act like Sparrows Point is their place. They're opinionated and narrow-minded and they let you know that they don't think you should be where they are. To their way of thinking, blacks aren't supposed to be successful."

The language of many white steelworkers reiterated the idea that black men "don't fit in," "don't want to work," "don't do well at the Point." A few white steelworkers had the insight to recognize that the presence of black men working in steel is perceived as a dangerous threat to job security. Tom Craig, a young white man who worked at the Point as a machinist, expressed his disapproval of the hostility toward African American steelworkers with the story of a black steelworker who showed up at a party given at the home of a white coworker: "All night long the white guys were hollering, 'Hey, Mike, we're going to get you a lantern and make you stand in front of the house.' It was good-natured kidding, but with a meaning behind it. There's nothing they would like more than to have Mike standing out front with a lantern—to have blacks be in a certain place beneath the white man." Most white steelworkers interviewed in the 1980s steadfastly denied any discrimination against blacks at the Point, insisting that blacks are treated equitably, "as long as they are willing to work," implying that any inequities must be the result of black incompetence or laziness.[28]

Black steelworkers who were interviewed in the 1980s were matter-of-fact about the ways in which racism affects their work lives, and a belief that "generally there is always a way to get through." Some older black steelworkers declined to be interviewed, indicating that they preferred to forget their humiliating experiences. The African American steelworkers I did sit down with each had a strategy, like Peter Wallace, who spent twenty-five years in the pipe mill:

> You have to get yourself right about dealing with the racism. Some black men stay to themselves and don't talk to any of the white

> guys. Me, I say hello to everyone; I'm friendly and most guys will eventually come around to accepting me. But whites don't socialize with blacks so you put yourself somewhere else. White guys will keep you from being a real part of their little groups, so we learn how to have our own fun, playing cards and telling stories, helping each other out, using street names that only certain people know.

The black culture that developed within the Sparrows Point steel mill and in African American residential communities of steelworkers included both formal and informal black associations. Ron Moore relished the strong bonds he made with other black workers:

> When I worked in the old cindering plant it was a predominantly black department and the guys bonded like a family. One of the guys taught me how to drive and took me to get my driver's license. When my mother died, all the guys were there . . . 30-some years later. We worked swing shift, and when we worked 3 to 11, after the shift we would go bowling or go to a nightclub. When we worked 11 to 7 we'd get off in the morning and we'd take turns having breakfast at each other's house. We formed a bond.

At the Point, black men developed strong social, cultural, religious, and political structures that were the source of pride, independence, and fellowship as well as a protection from racial tension.[29]

Racial tensions at the Point affected all steelworkers' families. White men brought home to their families the residue of either their own hostility toward black coworkers or the stress of feeling powerless to change or avoid the unmitigated racial tension that was part of the work environment. Racial antagonism at the Point also took a toll on white families who experienced a breach between steelworker husbands who had problems with blacks and wives or children in the same family who often were more tolerant. African American steelworkers brought home to their families the accumulated stress of having to get themselves "right about dealing with racism." Some black steelworkers suppressed their anger with alcohol or sullen withdrawal, while others were busy in the churches or recreation leagues in their communities. Ron Moore, a thirty-year-veteran at the Point, described why his father, who had also been a steelworker at the Sparrows Point steel mill, maintained an emotional distance from his family that was characteristic of a whole generation of black men working low-paying jobs in steel in the 1940s, 1950s, and 1960s:

> They came in here, and if they were black they couldn't eat the food in the restaurant; they had to go around the back way and get a bag. They had to deal with that. No matter how smart you were, you had a shovel. That's very stressful, because you know there's a white guy, he comes in and you train this white guy to do your job and he becomes your boss. That was something they had to deal with and they brought that stress home on some heavy shoulders.

The toll on the families of black steelworkers can be calculated in part by the accounts of their sons and daughters.[30]

It was only the Consent Decree of 1974 that successfully challenged job discrimination based on race at Bethlehem Steel. In Baltimore the Steelworkers for Equality, an organization of black steelworkers, joined a nationwide civil suit charging a pattern of racial discrimination throughout the steel industry. When Bethlehem Steel joined with the other big ten steel companies in agreeing to change practices in hiring, promotion, and seniority, there was a significant backlash. White steelworkers stood the accusation of discrimination on its head, complaining that lost opportunities for promotions for whites were the result of the Consent Decree and the consequent favoritism to blacks.[31]

Ironically, deindustrialization has coincided with a significant improvement in race relations for those men who still have jobs at the Point and who tend to be younger and more skilled. Today it is harder to generalize about race relations at the Point, because modernization has resulted in the recruitment of younger workers with high-tech skills who bring with them the perspective of a generation formed by multicultural social and educational experiences. In 2001 Norman J. Brown Jr., a young African American steelworker who came to the Point from the Navy, wrote a poem protesting not race discrimination but the class biases of a news reporter: "What did you come here to see?/Men wearing dirty clothes—unlearned and unable to hold intelligent conversation?/I am a high-tech individual who makes up a new work force on the rise."[32]

Brown's poem motivated me to update my interviews with steelworkers, and in 2002 I asked a diverse group of people to read an earlier version of this chapter. These most recent interviews about work at the Point included accounts of changes in race relations, the acceptance of women steelwork-

FIG. 7 Until Steelworkers for Equality filed a discrimination suit in the 1960s, most African American men were relegated to the hottest, dirtiest, and lowest-paying jobs at Sparrows Point, particularly those in the coke ovens.

ers, and improvements in the safety program. Unfortunately, retraining is the benefit most talked about since Sparrows Point is expected to lose even more jobs in the near future. Nonetheless, it is important to document the disclosures of dramatically altered interracial experiences at the Point that resulted from the most recent interviews.

Jim Kramer, a white electrician in his early 40s who has worked at the Point for twenty-six years, talked about a changed racial environment:

> I can definitely see the ways in which race relations have improved, even in my own work team. There was a black guy, back in the 80s we used to butt heads, and I think at lot of it was competition over jobs because we were both in the same electricians' apprenticeship. Back in the 80s our relationship was barely civil and when we got assigned a job, it was: "I don't want to work with Ron. I don't want to work with Jim." Today we are good friends. We like each other and we socialize together.

This level of genuine friendship between a black man and a white man was unknown in the 1980s. It is one of the ironies of deindustrialization that modernization and downsizing began eliminating thousands of well-paid jobs, just at the time when the Consent Decree brought together in the same departments and on the same work teams black men and white men who have commensurate skills and who are earning commensurate salaries.

Jim Kramer made a suggestion during his interview that I had never heard from a white steelworker in the 1980s: "This is my perspective as a white guy. I think you should talk to one of the black guys on my work team." Jim's consciousness that there are multiple perspectives on race was perhaps the most important indication of a change in racial sensitivity.

Jim's African American friend, Ron Moore, did, indeed, have a different perspective: "There was an incident just a few years ago where some white guys down here tied a dummy to a truck like the black man who was dragged to death in Texas. They thought it was funny. They got fired, but they got their jobs back. It's ugly, but you have to accept slow changes. There are still white men here who belong to the Klan and let you know it." In spite of some extreme incidents, Ron generally refers to white steelworkers who express overt racism in a tone that suggests that they are clumsy dinosaurs, culturally debilitated and unable to adapt: "You will still hear the 'n' word, not as much. There's some older guys, a certain type of die-hards. There's a guy named Johnny, I keep telling him, 'You're not allowed to use that word.' I don't think he means it personally, but he's been using it for so many years, it just comes naturally."

Ron Moore is primarily interested in talking about the ways in which the Consent Decree has opened the door for white and black steelworkers to work in the same departments day after day and, "start talking to one another until pretty soon you're asking about their kids, 'How's Tracy, how's Crystal?'" Ron saw personal benefits in his newly developed white friendships: "I feel really grateful, it has been a pleasure to be able to be around these white guys at work. They have taught me things that I didn't have exposure to before." He confirmed the importance of the close friendship between Jim Kramer and himself: "I feel fortunate that I have Jim as a friend. He came to my wedding, and that was very emotional for me, it meant a lot."[33]

Both Ron and Jim agreed that changes in the industrial system at their workplace, including improvements in safety and in race relations, had directly improved the quality of their family lives. For Ron, having white friends from work at his wedding was one step in removing the stress of workplace discord that he used to bring home. For Jim, having a black friend at the Point allows him to communicate more authentically with his wife and daughter, both of whom enjoy relationships with a diverse group of people in their work and school environments.

An ethos of manliness within a harsh work environment continues to characterize the Sparrows Point steel mill, embedded in an industrial work environment that is exhausting, dangerous, unhealthy, hierarchical, and palpably masculine in its combination of both camaraderie and combativeness. Working conditions and social relations have changed dramatically over a century, but the Sparrows Point steel mill, with its long hours, swing shifts, and brutal environment is a workplace that continues to have a profound effect on steelworkers' wives and families. It is also a work environment that is universally described by everyone who has ever worked there as "a man's world."

3

Boarders and the Long Turn in a Company Town: Sparrows Point Wives, 1887–1945

My grandmother would say, "You don't know what hard work is until you've worked in a boarding house." Because she was the one who woke the men up in the morning and got them ready for work, she would feed them breakfast and dinner and also pack their lunches for them. I can remember my grandmother talking about falling asleep in the bathtub because she would be so exhausted from working at the boarding house all day.

—CAROL PETERSON, APRIL 1993

The income-earning experiences of the wives of steelworkers at the Sparrows Point mill were unusual in comparison to the wives of other workers at the turn of the twentieth century. In 1900 only 18.8 percent of women were employed nationwide. The typical white female wage earner between 1900 and 1930 was a working-class daughter. During that period, between 85 and 88 percent of the daughters of workers sixteen or older were in the paid workforce.

Married white women generally did not work for wages outside of their homes in the pre–World War I period. Only 5.6 percent of all married women were at work in 1900, and by 1930 only 11.7 percent were wage earners. Married white women's occupational niche consisted of jobs that were done either in their homes or close to their homes so that their child care responsibilities could be managed. Some took in laundry or piece work, particularly from the garment or artificial flower industry. Some worked shucking oysters or in laundries or candy factories that were within a few blocks of their homes. Still others worked as janitors in offices and public buildings, jobs they could do at night when husbands would be at home to care for children. The small numbers of wives of white workers who took jobs outside of the home did so to avoid poverty.

During the same period, married black women were ten times more likely than married white women to be in the paid workforce. Of all employed black women, 92 percent were agricultural workers or domestic servants. By 1930, even after the large-scale migration of African Americans to northern industrial cities during World War I, 90 percent of all employed black women remained in the lowest-paying jobs in agriculture and domestic service. With few exceptions, higher-paying work was closed to black women.[1]

Steelworkers' wives, black or white, did not enter the all-male world of the Sparrows Point steel mill until their labor was required during World War II. Nonetheless, they worked incessantly to make the production of steel possible. The domestic activities of Sparrows Point wives followed a taxing regimen that included budgeting, shopping, cooking, childrearing, and managing complicated households. The 434 wives who sent their husbands into the steel mill each day in 1900 also packed lunches and dinners for a total of 366 adult sons who lived with their families and worked at the Point. In addition, at the turn of the twentieth century most Sparrows Point wives were taking in single male steelworkers as boarders, providing the unmarried steelworkers who boarded with families the domestic services that were essential to the productivity and profitability of the Maryland Steel Company. In many cases Sparrows Point wives took in such a large number of boarders that they were essentially running small boardinghouses that constituted home-centered businesses contributing significantly to the economy of the household.[2]

In June 1900, George S. Webster provided a rich inventory of the town of Sparrows Point, signing each of the thirty-five sheets of the Sparrows Point census with a flawless cursive style indicative of the deliberateness with which he recorded the composition of every household in this steel community that was just thirteen years old. Just as the manuscript census gives an accounting of the ethnicity, occupations, and regional origin of men who migrated to Sparrows Point to work in the mill, it also provides insights into the living and working conditions of Sparrows Point wives.[3]

According to the census, women first moved to Sparrows Point as the wives and adult daughters of men recruited to work in the mill. The women who lived in Sparrows Point in 1900 were predominantly married women living in households that included husbands, adult sons, and single male boarders leaving and returning from long turns in the mill, in addition to other children, and sometimes parents or in-laws. Their responsibilities for maintaining these households complemented those of men whose long hours in the mill, and periodic twenty-four-hour shifts, made it necessary for male steelworkers to rely on women for a full array of domestic services done without electricity or modern appliances.[4]

Because the labor requirements of the new mill attracted mainly single men, the 1900 census shows an overwhelmingly male population in the community, with 1,914 adult men outnumbering the 661 adult women three to one. In addition to the 434 husbands and the 366 adult sons who were living with their families, an extraordinary 65 percent of Sparrows Point's male population lived in boardinghouses, hotels, shanties, or boarded with families. Of this group of 1,006 single men, which was the largest demographic group in the town in 1900, about half boarded with families while the other half were housed in all male, dormitory-style accommodations, whether hotels, large boardinghouses, or shanties. The cooks and waiters who prepared and served meals for steelworkers living in large boardinghouses were listed in the census and included African American women who were married, along with native-born white women who were either young and unmarried or widows. The few men who waited tables in the boardinghouses were young men and either immigrants or African Americans. For steelworkers who lived in shanties a mess hall provided food cooked by men employed by the company.[5]

The system of boarding in a family home was so widespread in Sparrows Point in 1900 that nearly three-quarters of all steelworkers' wives took in boarders to supplement the household income. Many Sparrows Point wives had large numbers of single men boarding in their homes. In the 73 percent of households that had boarders, the median number of boarders was five, and in some homes there were twice that number of boarders. The household of Daniel and Mary Wise, at 707 I Street in the African American community, included three adult sons, ages twenty-two, twenty, and sixteen, and eleven single male boarders, all of whom worked in the mill. The oldest son was married, so Mary Wise had her daughter-in-law as well as an eighteen-year-old female servant to help with the laundry and food preparation for the fifteen steelworkers living in the family home. By all accounts, Mrs. Wise was managing a crowded boarding establishment with only two women to assist her.[6]

In the census records, a household headed by a steelworker husband was never identified as a boardinghouse, regardless of how many boarders lived in the house. The total number of boarders in a residence did not determine whether or not the household was identified in the 1900 census as a boardinghouse. It was the husband's status as the head of household that determined the economic designation connected with that residence. A household headed by a man was consistently identified by the occupation of the husband in the mill, while the woman to whom he was married was identified as "wife" even if she was managing a household with ten or more boarders. This meant that the husband, not his wife, received a designated economic role.[7]

In only one case did Webster identify a household with as few as five boarders as a boardinghouse. Significantly, that household did not have a male head of household, but was headed by a widow. In addition to three adult sons, this widow boarded five single men, a considerably smaller number of boarders than many households that were not identified as boardinghouses. For the purposes of the census record, Webster conflated the designation of wife with domestic service. A wife, according to Webster's reckoning, could never have an occupation other than "wife" no matter how many boarders were paying rent in the household that she managed.

One of the largest boardinghouses in the company town was located on East E Street where sixty-eight men boarded. A white couple in their thirties ran this boardinghouse with the help of four single white women who worked as waiters and two single African American males who worked as the cook and the dishwasher, all of whom boarded at the facility. Another boardinghouse, the Patapsco Hotel, boarded sixty-one men and two women, including ten hotel employees and fifty-three men who worked in the mill and the shipyard. The head of the facility was a twenty-nine-year-old man who supervised a sizable staff, including a cook, a baker, and two chambermaids. Five African American men in their late teens and early twenties who had migrated from Virginia and North Carolina were also living and working in the hotel as waiters.

These two large-scale establishments were performing the same functions as the smaller boardinghouses that Sparrows Point wives ran in their homes. In every case, domestic tasks that were performed for boarders included washing clothes, shopping, cooking, serving food, cleaning, collecting rents, and budgeting. Managing these small businesses in their homes required that Sparrows Point wives have complex management skills, the ability to collect rents and establish rules, and the strength and endurance to perform substantial domestic services on a daily basis.

Sparrows Point wives ran these businesses in conjunction with raising children, a pattern that makes the boarding system even more impressive as a business and management achievement. In 1900 Sparrows Point was a town that was just thirteen years old. Its residents were overwhelmingly young people, either couples or single men in their twenties and early thirties, ages when people found it easiest to pick up stakes and migrate. Among this relatively young population, the median age of Sparrows Point wives was thirty-three. Since most couples were at the beginning or in the middle of their childbearing years, the size of families was typically small. The median number of children for both black and white Sparrows Point mothers was three, and forty-six couples in their early twenties had no children in 1900, although undoubtedly those families would grow over the next two decades.[8]

Sparrows Point families suffered from the high child mortality rate characteristic of the late nineteenth century. Of the 434 wives, an astounding 174, fully 40 percent of the wives in the community, had buried at least one child. In some families the deaths of children was unremitting. Elizabeth Barrett was a white woman who was forty-six in 1900 and had been married for twenty-seven years to fifty-two-year-old Carney Barrett, a blacksmith in the mill. Elizabeth had given birth to 16 children, and 5 of those children, ranging in ages from six to twenty-one, were living in the Barrett household in 1900. The census indicates that 9 of the Barrett's 16 children had died.[9]

African American families also experienced high rates of infant mortality. Of the 26 black couples living in Sparrows Point in 1900, 20 couples had had children, and 9 of those couples had buried children. Lottie Canady, age twenty-three, was married to a fireman in the rail mill, twenty-seven-year old Abe Canady, and lived with him at 705 I Street. Lottie and Abe Canady had been married for four years, had four children, and had buried all of them. Their neighbors, Luther and Betsy Grant, who lived at 711 I Street, had been married for thirteen years and had buried 4 of their 6 children.

Only a handful of single women lived in Sparrows Point in 1900. Unlike Sparrows Point wives, single women could not be fitted into the category of wife, so they were assigned occupations in the census record as servants, seamstresses, or laundresses. In addition, four unmarried daughters of skilled workers who lived with their families were listed in the census as clerks in the company store. Depending on their ages, adult daughters were responsible for varying degrees of domestic work within their own families, but like their mothers they had no occupation assigned.

There were other gendered dimensions to the Sparrows Point community, and none was more striking than the educational system, which was considered highly advanced for the turn of the twentieth century. When Rufus Wood designed the village where steelworkers were to be housed, educated, and serviced by company-owned or -sponsored stores, churches, and fire and police forces, he also had in mind an elaborate plan for molding desirable workmen who would be, according to Maryland Steel documents, "sober," "industrious," "reliable," "steady," and "willing to work." A report on the company school, written in 1907, announced proudly that through its vocational education, "older boys are so thoroughly fitted in the use of the wood-worker's tools that when graduated, they are placed on the roll of employees to take places when men drop out." Even more revealing of the conventions of gender in this steelmaking community is the report's much more detailed outline of female education:

> The idea at Sparrows Point is not only to mould the girls into women but home planners and homemakers. Even girls ten years old go into the training kitchen, don their aprons, roll up their sleeves, and do such things as knead dough for bread and biscuits, make coffee that is clear and fragrant. On her little one-burner alcohol stove, a girl cooks a breakfast of oatmeal, omelet or bacon and eggs, perhaps biscuits or toast. . . . A wife who is a good cook is a "joy forever" to her husband, though she may not be a "thing of beauty." . . . The girls are shown how to use the broom and duster in a way that will not strain their backs. A model laundry with stationery [*sic*] tubs gives them practice in washing so they can tell how clothing should be washed without tearing.[10]

The founders of the Sparrows Point mill intended to cultivate among young girls the domestic skills that were necessary to maintain efficient households. In the early 1900s, without modern housekeeping equipment, Sparrows Point wives were working under labor-intensive conditions, using simple laundry tubs that had to be filled with water heated on the same one-burner alcohol stoves that were used for cooking. Frugal, well-organized households were an essential complement to the organization of work in a steel mill, where men earned wages that were considerably higher than farm labor but which provided modest incomes during the nonunion era. From that perspective the educational system in Sparrows Point was designed to foster standards of both female and male roles that would be well suited to the demands of profitable steelmaking.[11]

Interviews with individuals who lived in Sparrows Point, as well as with their children, grandchildren, and great-grandchildren, allowed me to paint a portrait of the Sparrows Point community after 1900. When asked about life in Sparrows Point during the first half of the twentieth century, the descendents of Sparrows Point wives and mothers described domestic work before World War II as a life of drudgery. In the first two decades of the century, the absence of household conveniences meant that women spent long hours doing routine household chores that required their full-time attention. After World War I, electricity was installed throughout the town of Sparrows Point, but even that major improvement didn't alleviate the constant toil of steelworkers' wives who were raising large families.[12]

Scrubbing industrial debris off floors, windowsills, and furniture, and washing clothes, bed sheets, and curtains were made particularly arduous because the town of Sparrows Point was immediately adjacent to the mill, and smoke and dust saturated everything in the vicinity. Helen Reed described the town as she saw it as a child, covered in industrial residue: "I remember walking through Sparrows Point and everything was orange. The streets were orange, they had that rust orange color. The buildings were orange, the cars were orange, even the houses were orange. My grandfather's work clothes all had that rust orange color that wouldn't come out." A writer for the *Baltimore Sun* was incredulous when, in 1906, he reported the juxtaposition of a charming tree-lined town enveloped in industrial pollution: "Over the crest of a small mountain of iron ore, a dozen chimneys give forth enormous clouds of smoke—red, yellow, brown, green gray, white and black. Across an open plain of grotesque, distorted lava, there basks a quiet little town—a town of green, shady streets and gardens full of roses." Middle-class journalists—as well as the steelworkers themselves—saw the pollution from the mill from the perspective of industrial prosperity and the jobs they provided.[13]

The Sparrows Point wives who kept the houses in the "quiet little town" sparkling clean were not so rhapsodic about the industrial pollutants that they battled on a daily basis. Reed reported her mother's irritation at the

FIG. 8 Men working in the strip and tin reheat furnaces in the 1920s brought work clothes home covered with grease, dust, and grime, and it was the work of Sparrows Point wives to launder these clothes week after week.

laborious cleaning chores caused by the nearby mill: "Like the other wives, she would test the wind before she hung the laundry out to make sure it was blowing in the right direction so she wouldn't have orange soot all over the laundry. My mother used to complain and complain about the red dust and the black soot on the table, and my father said, 'The day we don't have to wipe the table, people will be hungry!'"

Sparrows Point wives maintained high standards as fastidious housekeepers, and it was a sign of competence and respectability to maintain a clean home. This was an endless task with smoke and dust gushing from the blast furnaces and the coke ovens. One resident of Sparrows Point commented that, "when it came to washing and hanging out clothes, you'd have to look to see which way the wind blew. If it came from the open hearth, you were going to get red clothes. . . . If it came from the east side, the clothes would be black. If you had a lovely high windy day, there would be people washing all over town."

Daunting as the ubiquitous soot and smoke were, Willa Martin remembers that until the 1920s her grandmother battled the red dust in Sparrows Point without electricity or indoor plumbing:

> When my mother was a small child my grandparents moved to the 600 block of E Street, which was a step up. They were able to move up to what my mother considered a really modern house. It had indoor plumbing for the first time, a telephone for the first time, and they had electricity. Prior to that my grandmother had raised her nine children without electricity or indoor plumbing.

This was the kind of life that required that women work without respite, their labor controlled by the ceaseless nature of the tasks in much the same way that their husbands' work inside the mill was controlled by the time clock.[14]

Martin is proud of her grandmother for coping with cramped physical space and the absence of luxuries or privacy: "My grandmother raised her nine children in a house with one bathroom. They lived in one of the better houses, and they did have large rooms, a living room, a dining room, and a large kitchen. My mother remembers sleeping three in a bed when she was growing up." This respect for women who can accommodate arduous work in challenging circumstances is a widely held value in Dundalk and Turner Station, and contemporary women from those communities talk about themselves as being a part of a tradition of women who are competent and able to work hard.[15]

The household tasks of Sparrows Point wives followed rhythms that paralleled the ways in which shift work determined the organization of the community and the rhythms of family life. According to Margo Lukas:

> On Mondays my grandmother would wash clothes. The washer was in the kitchen or on the back porch, and the family would get leftovers or something small to eat because Mom was washing and it was an all-day into the night task, even though they didn't change their clothes daily as we do nowadays. They were the old "get a bath on Saturday night" kind of kids, because there were nine kids and five beds full of bed clothing.

By moving to a house with electricity, Martin's grandmother gained an advantage over the wives who appear in the 1900 census because she had the help of newly available household appliances like the washing machine. Nonetheless, with nine children in the family just doing the laundry took working the entire day and into the night.[16]

Meals were determined by shift work schedules. Ordinarily, supper was served when Willa Martin's "Pop" got home, but the double shift—even after labor reform eliminated the twenty-four-hour shift—meant that once every two weeks steelworkers were away from their families for sixteen hours:

> My grandfather worked in the open hearth and they shoveled all the stuff in by hand. Every other week he would go to work on a Sunday afternoon and he would work a double shift from 3:00 on Sunday afternoon until 7:00 Monday morning. Sunday evening, Mom would pack his supper. I can still see it; he would have this oval aluminum lunch pail. They would put hot water in the bottom and then there was a tray on top, and she would have dinner, potatoes and meat or whatever, and she would put it in there, and Mom would send us over with Pop's supper.

When Pop had to work a double shift, his wife prepared a dinner that had to be carried into the mill. A woman entering a steel mill in those days was taboo because it was believed that a woman's presence could cause an accident or even a steelworker's death, so one of the children delivered the dinner pail. Dundalk historian Ben Womer also remembered that in the 1920s when his father was working the long turn on a Sunday, his mother prepared dinner and packed it in a big aluminum bucket, and Ben carried his father's dinner into the open-hearth department and brought it right up onto the floor.[17]

When I asked Willa Martin about the lives of women who lived in Sparrows Point during the first half of the twentieth century, she responded with both admiration and sadness. During the nonunion era prior to 1941, life in a steelmaking community was one of relentless work for both men and women. Martin remembers with veneration this heritage of hard work, but she is realistic about the deprivations inherent in the life of a steelworker's wife at the beginning of the twentieth century: "It was just a life of work and a seriousness there, with the belief that fun things were kind of foolish things. I never remember my grandmother doing anything but working. She had a beautiful yard and she would go out there in the evening after working all day long taking care of that big bunch of people. And then she'd work some more, planting flowers."

In the Sparrows Point families that went on to have large families, child-bearing was such a regular occurrence that it is the routine of it that Martin remembered, rather than the birth of an individual child being a special, memorable event:

> My mother tells the story of my older aunts minding her when she was a child, and telling her: "Come on upstairs, Mom has another baby." My older aunts saw all these babies, but my mother was only witness to two after herself. My mother often told me, "I can remember us little kids going in there to see the baby." And, of course, one of the older girls would have to take over the household while their mother had that baby. All those families lived the same way. That was just the way life was.

Older daughters became a part of the domestic support system for a household as soon as they reached the age when they could be a "mother's helper." In much the same way that Rufus Wood had planned for female education in the 1890s, through much of the twentieth century daughters of steelworkers were trained to be conscientious wives and efficient household managers who were also practiced in the routines of child care.[18]

When asked about her grandmother's role in the family, Martin recalled a strong woman who was boss of her domestic domain, budgeted the household money, and controlled the behavior of the children:

> My grandmother had authority in the family. She had the authority. I think it was a situation where probably for all those many years my grandfather considered himself the head of the household, but I'm going to tell you something, she was the boss. When it came down to decisions I guess the two of them made the major decisions, but if the children wanted anything, they went to their mother. And she was very much, very definitely the boss.

Sparrows Point wives were also involved in managing the financial affairs of their households, especially by allocating the wages earned by male steelworkers. Initially, this pay was rendered in the form of a credit system that gave each steelworker a charge number, and on payday his wages were credited to that account. Regular debits were made for the rent on the family

FIG. 9 Sparrows Point provided opportunities for young couples seeking economic improvement, c. 1900.

home, and beyond that the charge number could be used to purchase items from the company store. By all accounts, Sparrows Point wives were responsible for the decisions that concerned buying essential items like food and household necessities, and for budgeting for clothing and shoes for their husbands, their children, and for themselves. Finally, most husbands and wives participated jointly in decisions to buy luxury items like new furniture, rugs, and draperies. Moving to a larger house, which meant an increase in rent, was almost always made by mutual agreement.[19]

In the households that had boarders, an additional financial responsibility was collecting the rent. Based on the interviews with daughters and granddaughters of women who took in boarders between 1920 and 1945, collecting the rent was sometimes done by Sparrows Point wives and sometimes by their husbands. Husbands could more comfortably ask for rent from roomers, boarders, and male relatives, all of whom worked in the mill together. But once collected, the rent money was necessarily turned over to wives, whose responsibility it was to budget for and then purchase food and household supplies.

The system of boarding at Sparrows Point was a compelling draw for single men from out of town who needed jobs. In a letter sent from the wife of a steelworker living in Sparrows Point back to her friend Mill Kurtz in rural Pennsylvania on March 12, 1934, Charles Kurtz was encouraged to seek a job at the Point and told that he would have no trouble finding a place to stay:

> If he gets to work he won't have a hard time getting room & Board because the Co. owns everything in this place and they have women that run boarding houses and their [*sic*] awfully nice and if he wouldn't have enough money to pay it right away the Co. would just take it off of his first pay. And if he get on for Mr. McNear he'd probly [*sic*] take him to his place to eat, for his wife just serves meals and then you room some other place.[20]

Thus, wives of steelworkers even negotiated employment for friends and loved ones.

Before 1945, the steelworker wife at Sparrows Point may have been the primary financial manager of the household, and she may have been the significant authority in the lives of her children and grandchildren, but her domain was a limited one. The company owned and maintained repairs on her home, while her husband earned the family income inside a mill that she probably had never entered. Nonetheless, through the first half of the twentieth century, Sparrows Point wives, particularly those who took in boarders, had a significant role in providing for the economic well-being of their families and providing for the maintenance of the labor force for the Sparrows Point steel mill.[21]

Through World War II, Sparrows Point wives continued to contribute to the household income by renting rooms in their private homes to single men who came to Sparrows Point for jobs in the steel mill. These men paid rent either as boarders who both slept and ate with a family, or as roomers, who rented a room with a family but took their meals in a restaurant or boardinghouse.[22]

Peggy Lipsett remembered a whole series of boarders who lived in her family's home while she was growing up in the 1930s and 1940s:

> We had boarders in our Sparrows Point home for at least seven years. Some of the boarders stayed with us for quite a while and I

> remember them well. We had three bedrooms upstairs and one was a right small room, and my sister and I slept in that room and Mom and Dad had a bedroom downstairs where the dining room should have been. The other two rooms were occupied by boarders.

Peggy remembered that the income earned from taking in boarders and lodgers was so important to the household budget that families made sacrifices of their own space and privacy in order to make room for the single men who shared their homes: "We only had one bathroom. Four roomers and four house residents and we only had one bathroom. The only way this worked was that the roomers got up in the morning before we did and bathed before they went to work." Especially during the 1930s, income from boarders was often the ballast for steelworker families.[23]

Sparrows Point wives who were running households with boarders played an important role in the lives of the thousands of single men of all ages who came to the Point without families in order to earn the money that could be made in industrial employment. Some single steelworkers were quite far away from their families, determined to earn enough money to return home and begin a farm or business enterprise of their own. Lipsett tells stories of single men working at the Point who were hundreds or even thousands of miles away from their families and living quite lonely lives: "These roomers were not teenagers. They were men in their twenties and they lived as single men, just working at the Point. It was a sad situation for these roomers in a way. A lot of these people didn't have any industry where they lived, and some of them, the whole time they lived at the Point they were looking forward to going home."

Lipsett's mother was typical of Sparrows Point wives who cooked and cleaned for single steelworkers, and who also were a mature female presence in the lives of young men who spent long hours in an all-male work environment. Wives who contributed to the household income by taking in boarders also cared for some of the emotional needs of lonely young men through the comfortable homes, nourishing food, and daily kindnesses they provided:

> Two boys had come to the Point from Nebraska and the youngest one, Joey, didn't like it at the Point because we were a big city to him and he liked those open spaces. They saved their money and eventually they moved back to Nebraska and married and probably bought a farm or started a small business. They appreciated what

> my mother had done for them all that time, and after they left they sent her a Christmas card every year.

Lipsett's mother provided a home-cooked meal and the warmth of her home, making it easier for the young men from Nebraska to endure the separation from their own families.[24]

Through the World War II era, Sparrows Point wives who were widowed by mill accidents took jobs in restaurants, stores, cafeterias, and large boardinghouses that were owned by Bethlehem Steel. Linda Hartmann's mother was a widow who assumed the daunting occupation of managing one of the boardinghouses that accommodated as many as thirty or forty single male steelworkers: "In about 1943 my mother was widowed. First she worked in the cafeteria at the main office and then she took over a huge boardinghouse for about eight years. This was not an easy job for a woman to handle, but she didn't have a choice because she had to support her family after her husband died and she had to make a go of it." Unlike the configuration of women's economic roles in 1900, by 1940 there were several women managing boardinghouses outside their homes.[25]

In addition to yard work, household cleaning, and the laundering of towels and bed linens, the work involved preparing breakfast, lunch, and dinner for two or three dozen men: "She had a daughter and her two other sons were in their teens and all of them helped with the chores. Her daughter helped with food and laundry and her boys helped too, with everything including peeling the potatoes. The boys took care of the lawn, which was large because it was such a big boardinghouse." Managing a boardinghouse involved major responsibilities for all members of the family who were not yet in the paid workforce.[26]

Running a large boardinghouse constituted an even larger business enterprise than did taking boarders into the family home. Hartmann recalled the quantities of groceries, meals, and linens her mother managed on a daily basis: "She would buy huge fifty-pound bags of potatoes. This was after the company store had closed so she relied on her sons to go with her into Dundalk or over to Eastern Avenue to buy groceries. She drove a car so she didn't have to rely on the streetcar." As the manager of a large boardinghouse, Hartman's mother crossed a significant gender barrier by virtue of the fact that she owned and drove an automobile. In the 1940s husbands bought and drove the family car, and with rare exceptions the pattern of women going through life as passengers continued well into the 1970s and 1980s.[27]

A woman managing a boardinghouse in Sparrows Point also had to hire and supervise employees, usually adult women who were also widowed:

> She had to hire women to help her with the preparation of the meals and with changing the bed linen. She usually hired widows, and there was one woman who worked for her for a long time. She never hired young people to work in the boarding house because it was just a temptation to the men. Maybe she tried it once and it didn't work out. I never remember any of the black women working in the boarding house, although black women did work doing ironing for many of the white families on the Point.

Young women were disqualified from this boardinghouse job because of the potential of romantic alliances with the boarders. Thus, the woman running a boardinghouse had to demonstrate both discretion and keen judgment in managing her staff, even if, or perhaps especially if, that staff included her own daughters and sons.[28]

Hartmann's mother also had an additional responsibility if a boarder had to be evicted, which in most cases was because of the old nemesis that Rufus Wood railed against—alcohol abuse:

> A couple of times my mother had to deal with men in her boardinghouse who went off the Point and drank and had to be evicted from the boardinghouse. They were just not desirable tenants because they either drank or they were so loud or argumentative that they caused a row, and several times she had to get rid of men. Sparrows Point had its own police force so if she had a problem with any of the men she just had to call and the police would make sure the men got off the Point.

The policing functions of a company town like Sparrows Point made it easier to control unruly tenants, making it possible for women to run businesses and enter occupations that might have been difficult or impossible in a less controlled environment.[29]

Hartmann remembers that her mother was pleased to be able to run a business and support her family, but that she was frustrated by the fact that she was never able to take any time off. There was no opportunity for a vacation because there was no one around to take over. Women who managed or worked in boardinghouses in Sparrows Point passed on to younger

FIG. 10 A young family in a formal photograph, c. 1900.

women in their families recollections of the fatigue that came with relentless hard work. Eighty years later the stories that Carol Peterson's grandmother told about toiling in a Sparrows Point boardinghouse are still very much alive in her granddaughter's memory: "My grandmother was able to go back to the boarding house and work after she got married, up until she had her second child, who was born in 1921. When she had her first child she would still go down there and help get the clothes washed and pack the lunches." Between the 1880s and World War II, three generations of Sparrows Point wives worked to exhaustion each day, before the advent of automatic washing machines and fast food restaurants, and before private living arrangements for single men were built as part of the war effort.[30]

A large number of single men sought jobs at Sparrows Point between 1887 and World War II. That fact, combined with the low wages those men earned prior to unionization, helped to create an industrial community in which significant numbers of both married black women and married white women became the managers of complex households composed of boarders and family members working in the mill. The fact that wives in both white and black households were contributing to the household income by managing small boardinghouses in their own homes made the Sparrows Point experience an unusual one for the turn of the twentieth century, because nationwide statistics indicate that in 1890 black women were ten times as likely to be in the paid workforce as were white women.[31]

The Sparrows Point steel mill employed only men and needed the labor of thousands of them. Between 1887 and World War II, this practice contributed to the creation of a gendered social world that included an important economic role for both white and black wives of steelworkers. It was a role that would continue until the post–World War II prosperity, a prosperity that included union wages for steelworkers and that made possible a middle-class lifestyle and new roles as homemakers for steelworkers' wives in Sparrows Point, Dundalk, and Turner Station.

4

The Family Works the Schedule: Steelworkers' Wives, 1945–1970

Shift work just wrecks your life, but it wrecks your family more, if you're an essential member. Most steelworkers have given up on being central members of their families. Their sense of responsibility to their families is earning money to buy the things their families need. The rest is up to their wives.

—TOM PORSINSKY, JANUARY 1993

In January 1958, Bethlehem Steel announced that its mill at Sparrows Point was the world's largest steel plant. During its heyday in the 1950s and 1960s, employment at Sparrows Point reached its peak of 30,000. Skillfully negotiated by the USWA, the wages of many steelworkers were sufficient to comfortably support a breadwinner/homemaker family structure. When the era of boardinghouses ended in the 1950s, the preponderance of steelworkers' wives for the first time since 1887 were not engaged in income-earning enterprises centered in their homes, and few of them were in the paid workforce outside the home. The wives of Sparrows Point steelworkers had assumed new roles as homemakers.[1]

By the 1950s, steelworker families were gradually moving out of the company town of Sparrows Point, and the steelworker communities of Turner Station and Dundalk were expanding and becoming more prosperous.

Steelworkers' wives had smaller families and more laborsaving devices, and they did not need to contribute to the household income by taking boarders into their homes. During that period of relative prosperity, the wives of skilled steelworkers enjoyed leisure, financial security, and a standard of living far beyond earlier generations of steelworkers' wives.[2]

The formation of a new gender identity in the 1950s and 1960s among steelworkers' wives in families in Turner Station and Dundalk was embraced in both white and black families. White women in Dundalk enjoyed the less burdensome life of being a full-time homemaker, which also helped to bestow middle-class status on their families. Margo Lukas, who grew up in Turner Station, described homemakers in Dundalk as well as Turner Station in the 1950s and 1960s: "The housewives, the stay-at-home wives of men who worked at Sparrows Point, for the most part were pretty comfortable women. They were home, and the number of kids they had dictated what kind of lifestyle they lived." For reasons that had to do with the history of married women in the paid workforce, a higher proportion of white women in Dundalk were able to take full advantage of the choice of a homemaker role.[3]

Married African American women living in Turner Station between 1950 and 1970 were, nonetheless, more likely to be employed outside their homes than were their white counterparts in Dundalk. Because residential segregation severely restricted the number of neighborhoods where African Americans could live, Turner Station became a diverse community with differences in class position and family patterns marking its population. In nationwide statistics for 1890, married black women were ten times more likely to be in the paid workforce than married white women, although the work they did was primarily poorly paid agricultural work or domestic service. By 1930 married black women were three times more likely to be in the paid workforce, because married black males were excluded from better-paying jobs and needed supplementary income. But beginning with World War II married white women began going into the paid workforce, initially making relatively modest entrees. By 1970 the ratio of married black women to married white women in the paid workforce was only 1.3 to 1.[4]

Like black women elsewhere, it was not unusual, nor was it considered contrary to community norms, for married women in Turner Station to work. Kathy Murray grew up in the 1950s and 1960s in a neighborhood in Turner Station where her personal experience was one of women leaving home to work: "Most of the women I knew worked outside the home. They were nurses, hospital workers, teachers, and domestics. In my own family all of the women worked, and it was just assumed that women would work."[5]

In addition to the fact that Turner Station accommodated people across the social class and occupational spectrum, there was a second reason why married black women in Turner were more likely to be in the paid workforce. As a group, black steelworkers received markedly lower wages than white steelworkers until the 1974 Consent Decree allowed them access to skilled jobs at the Point. When I asked African American steelworkers and their wives about the post–World War II prosperity, Pete Wallace acknowledged that his income was not sufficient to support his family single-handedly. In a pattern more typical for African Americans, Pete's household required two incomes to achieve middle-class standards: "My wife worked for Social Security, as did the wives of lots of black steelworkers. We had a good life for ourselves. If the man worked at the Point and his wife worked for Social Security, we thought we had it made."[6]

Despite the fact that it was more common for the wives of black steelworkers to be in the paid workforce than for the wives of white steelworkers, the breadwinner/homemaker family structure was widespread in both communities between 1950 and 1970. During these decades there were many African American families in Turner Station, and in Baltimore as well, that could rely on the income of a steelworker husband to support a single income, breadwinner/homemaker household.

While white and African American women married to steelworkers during the 1950s and 1960s confronted life circumstances that were distinctively different, this chapter emphasizes that there were many black women in Turner Station whose husbands' work in the steel mill provided the opportunity for them to gain parity with white breadwinner/homemaker households in Dundalk. When this group is the focus, there are clear similarities in the status of steelworker households in Turner Station and Dundalk. The common experiences of those wives of black and white steelworkers who took in boarders in the decades immediately before and after 1900, reemerged in another form of common experience as women in both communities became homemakers in the 1950s and 1960s.[7]

In contrast to Kathy Murray, Margo Lukas gave a description of growing up in Turner Station that parallels the family life of those same decades in white Dundalk: "I came from a block where most of the mothers stayed home. Most of the men worked at Sparrows Point, and the wives took care of everything that had to be done at home, including cut the grass, plant the flowers, make market, and have their dinner on the table when the men

FIG. 11 The family car was included in photos as a symbol of middle-class prosperity, c. 1940.

came home." Homemakers in Turner Station, like those in Dundalk, were admired for their comfortable lifestyles and their ability to buy good food for their families, nice furnishings for their homes, and attractive clothes for themselves. Jacinta Cole observed that in Turner Station a steelworker's wife was easy to identify: "Mostly they were very well groomed. The purses matched the shoes and the gloves matched the coats, and you could easily pick those people out." Women in Turner Station who stayed out of the paid workforce and were supported by steelworkers who could provide for their families on a single wage were identified as having both privilege and status by those married or single women in Turner Station who worked outside their homes.[8]

The wives of Sparrows Point steelworkers—in both the black and the white communities—who were full-time homemakers, assumed a unique role during the post–World War II era because they were not contributing income to family budgets. They also had labor-saving appliances that transformed their roles from that of household laborers doing debilitating drudgery.[9]

Wives of steelworkers still cooked, cleaned, and raised children. They assumed responsibility for shopping for the new consumer items that the families of steelworkers could now afford, products that added convenience as well as status to their households. These included luxury items like new furniture and home decorations, household appliances, and ready-made clothing for family members. Homemakers still had plenty of work, particularly while they were raising small children, and especially if their families were large. Nonetheless, their roles as full-time homemakers involved less of the intense labor done by pre–World War II generations of wives who took an entire day to do the laundry by hand. The homemaker role assumed by both white and black women was a mark of family financial success, and steelworkers of that generation commented with pride in their interviews that, "My wife doesn't have to work."[10]

It was during the post–World War II era that the families of Sparrows Point steelworkers began to identify themselves as middle class. When I started interviewing, my tentative focus was "working-class transformations," but the steelworker families I interviewed rejected the category "working-class" and insisted that their neighborhoods were middle class. They justified this category by referring to both their incomes and their lifestyles. According to their way of reckoning class position, both Turner Station and Dundalk between 1950 and 1970 were middle class by virtue of home ownership, discretionary money to buy luxury items, as well as the homemaker status of wives whose breadwinner husbands were the sole providers for their families. Standards of well-tended yards, regularly swept sidewalks and gutters, and orderly behavior were maintained by sets of neighbors who helped each other in emergencies and disciplined each other's children. In addition, there was widespread membership in churches and civic organizations, voluntary associations like the Boy Scouts and Girl Scouts, and recreational leagues for young people. During the decades when Bethlehem Steel provided thousands of jobs with union wages, breadwinner/homemaker families flourished in both the white and black communities adjacent to the Sparrows Point steel mill.[11]

The availability of high-paying steelmaking jobs in the decades after World War II is remembered as a great boon to this industrial community. This period is the Golden Age that many retirees in Turner Station and Dundalk recall as an era of abundant manufacturing work, home ownership, and cohesive neighborhoods. Nonetheless, easy access to jobs in steel also perpetuated a gendered social world that sent men into the mill immediately after high school, and sent women, especially white women, to the altar with few job skills and little access to paid employment. From both

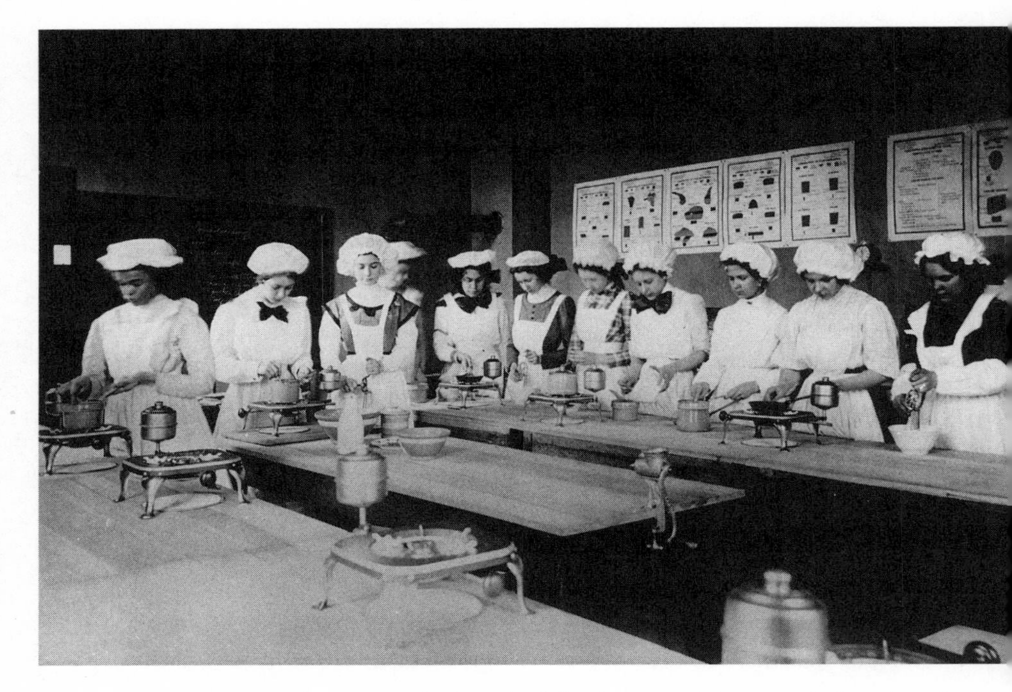

FIGS. 12A AND B From its beginnings in 1887, Sparrows Point as a whole reflected the same lines of segregation as the labor force in the steel mill. Although the women in these photographs all learned the same lessons of shopping and cooking on a limited budget, they could do so only in racially separated groups.

steelworkers and their wives, the Point claimed significant sacrifices. The testimony of men who have worked at the Point is filled with specific examples of the ways in which their jobs took a toll. The sacrifices made by wives of steelworkers has gone largely unrecognized, but can also be heard in the reflections of the women who negotiated roles that were deemed by community mores to be enviable—the homemaker wife of a well-paid industrial worker.[12]

In the course of their interviews, most steelworker husbands were quick to point out how hard shift work was on marriages and families. Larry Sayles described the pattern he experienced in his own household and that of his coworkers:

> For couples that wanted to build relationships that were healthy, shift work was always a challenge. He comes home at 11:00 P.M. with energy to burn. She has had a bruising day at home, wants to sleep. They try to share some responsibilities, but when he's work-

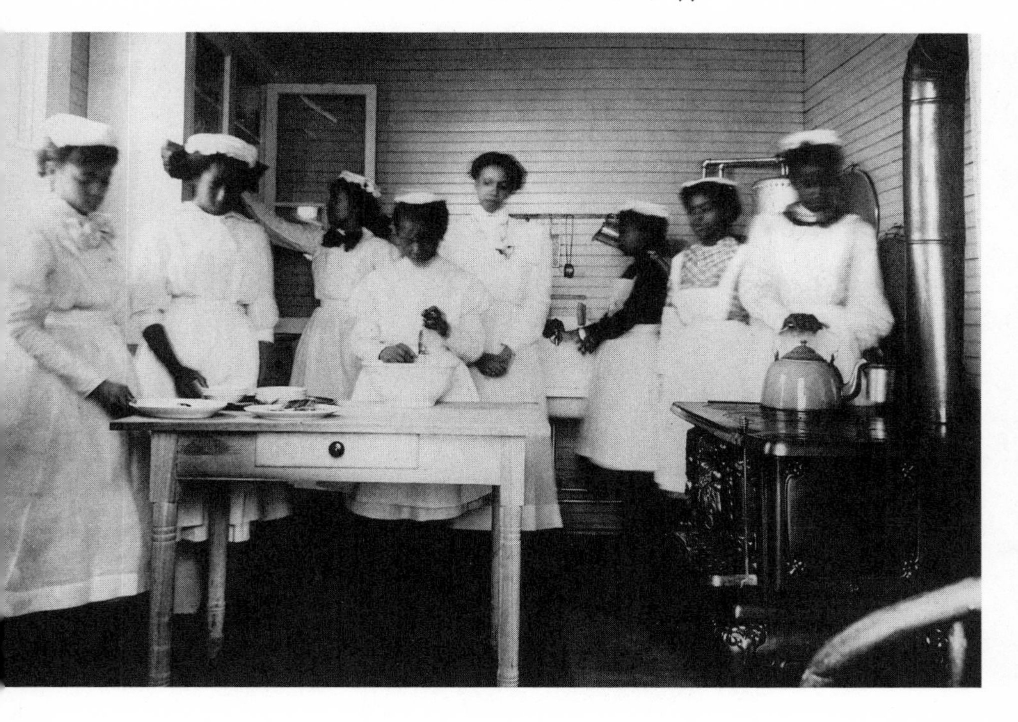

> ing three-to-eleven, most of the work of ferrying kids around, and doing the errands, falls on her. The school activities are strictly her responsibility.

Like most steelworkers who were interviewed, Larry was aware of the tension between family life and shift work.[13]

In the 1950s and 1960s, the availability of abundant overtime also conflicted with family cohesiveness. Overtime paid time and a half, meaning that the more a steelworker was absent from home, the more prosperous his family would be. At the turn of the twentieth century, the eleven- to fourteen-hour shifts in the steel mill dictated that steelworkers were either working or sleeping, but even the shorter shifts that were mandated for the steel industry in the 1920s did little to make steelworker husbands available as partners in home-centered activities and tasks. By the 1940s, steelworkers were seduced by highly paid overtime into working hours that were sometimes even longer than the unregulated shifts prior to the 1920s. Abundant overtime resulted in the fatter paychecks that enhanced the reputation of Bethlehem Steel as a desirable employer, but Larry Sayles explained how it also kept steelworkers away from their families: "Since many

FIG. 13 Mothers passed on to their daughters their aspirations to be married to steelworkers, which meant the opportunity to stay home, raise children, and care for the house and garden.

of the straight shift jobs, like seven-to-three, were also the lowest paying, spouses often were forced to trade family normalcy for money and upward mobility." The extreme inconvenience of well-paid shift work was the clearest example of how the organization of steelmaking served as a constraint on those steelworkers' wives who might otherwise have sought access to the full-time paid workforce.[14]

Bob Shaw saw that work outside the home was financially impractical for a woman whose husband was working swing shifts: "Since overtime paid so much more than most women were capable of making, many women married to steelworkers stayed home within the limited horizons of the block, the school, and the market." The tasks involved in fixing meals, getting children ready for school and caring for them when they were not, and maintaining a family life, fell upon the wives of steelworkers as a full-time occupation. The combination of her husband's erratic schedule and his ability to earn high wages colluded to keep the wife of a steelworker homebound, out of the paid workforce, and economically dependent. Swing shifts imposed burdens that wives and children had to accommodate.[15]

Wives and daughters of steelworkers who were interviewed about the quality of life for homemakers in the 1950s and 1960s in communities near the Sparrows Point steel mill do not extol this alleged Golden Age as an era that was without difficulties for steelworkers' wives. Grace Caldwell recalled how the accommodation of her husband's shift work schedule put substantial strains on their household: "Oh, God! It was the pits! He would swing every week between three shifts: seven-to-three, three-to-eleven, and eleven-to-seven."[16]

Theresa Porsinsky stated flatly, "It's a very difficult life. Shift work means that I can never count on my husband being here at a certain time." Theresa went on to explain that although she had gone her entire life without entering the gates of the mill, as the wife of a steelworker she was not insulated from the harsh working conditions that accompany the production of steel. The work regimen of changing shifts kept the production of steel going twenty-four hours a day, seven days a week, but the efficiency of the steel operation took an enormous toll on the personal well-being of the Porsinsky family. Theresa described herself and other steelworkers' wives as "prisoners of the swing shift."[17]

Even steelworkers who worked straight daylight shifts spent excessively long hours at the Point because a boom in production would force them to stay until their work was done. Sandra Murray remembered her steelworker father as overworked and frequently absent:

> When I was growing up my father's work at the Point kept him away a lot. There was just so much pressure on him that he'd come home and he really didn't want to talk to anybody, and not because he was angry, it was just because the pressure was so much. He'd go in at four o'clock in the morning and get home at eight o'clock at night. He would just pretty much come home, take a bath, sit down for about an hour to try to unwind, go to bed, and he'd be up and out before I even got up.

The erratic schedules resulting from unpredictable sales and production in steel were yet another constraint on steelworkers' wives.[18]

Wives whose husbands were working night shifts had the difficult task of keeping children quiet while steelworkers slept during the day. Gloria Robinson complained about the burden that shift work put on raising children: "When my husband worked nights I'd have to try to keep the kids real quiet while he slept in the back bedroom." Adult children of steelworkers recall growing up eating in the middle of the afternoon and tiptoeing around the house so that they would not wake up their fathers. Joan Zagorski reflected that: "The thing I remember most vividly was my father taking naps on the couch after his night shift, and everyone was supposed to be quiet."[19]

Wives of steelworkers also assumed the obligation of having food prepared for hardworking steelworker husbands. Gloria Robinson is an example of a wife who was particularly conscientious about this: "No matter when the shift ended I'd have a hot meal waiting for my husband, even if it was eleven o'clock at night." Sacrifices made in an exceptionally masculine work arena meant that steelworkers who were breadwinners claimed the right to make demands on the time and services of their wives, the most fundamental of which was to require a hearty meal when he came home from work. In a breadwinner/homemaker marriage, a steelworker could also make claims on his wife's leisure time activities and choice of companions, making it difficult for women to achieve a semblance of autonomy within their marriages.[20]

Pauline Pearson was still bitter twenty years later that her mother did few things outside the home for her own pleasure. One of those small personal pleasures was bowling, and her father begrudged his wife even that time away from household responsibilities: "The only thing that woman did was go bowling on Wednesday morning. And that night we would have hot dogs and my father would complain, 'If she didn't bowl, we'd have a real meal.'" Bowling was perceived by this steelworker/breadwinner as interfering with one of his wife's primary responsibilities—the preparation of a suitable evening meal that he could enjoy after a hard day at work.[21]

Bowling also extended beyond the conventional expectation that women would socialize within the family circle. Turner Station and Dundalk women remember that holidays and birthdays were always celebrated with family gatherings, and also that their mothers rarely developed friendships outside of the family or neighborhood except through church. Bowling included socializing with women who were neither neighbors nor family members, and it therefore was a suspect form of recreation.[22]

Long hours and changing shifts left many steelworkers chronically exhausted and excluded from a central role in family life. These arduous conditions often fostered drinking after work as one of the few accessible ways to relax after a shift, and a pervasive bar culture encouraged drinking as part of a masculine ritual that was obligatory for steelworkers. The effect on family life of this part of the work regimen of steelmaking was to make steelworkers even less accessible to their family members. Eric Logan, a former steelworker who had worked at the Point for eighteen years before being laid off in 1984, described his own devotion to "drinking with the guys": "What my kids remember of me when I worked at the Point was a drunken father who would stumble into the house at all hours. When they left out in the morning I was in bed. It was hard on family life because most of the guys drink a lot, and I joined the crowd because it was the macho thing." Interviews with the adult children of steelworkers from the Point include many accounts of devoted family men with little interest in drinking, but throughout the history of the mill thousands of men have turned to alcohol as one of the customary and habitual ways to relax after a shift. Participation in family life could not compete with a work culture that included constantly changing shifts and obligatory drinking.[23]

In discussing the family dynamics in the breadwinner/homemaker families within which they grew up, many women recall their fathers as hardworking breadwinners who expected an array of entitlements because of their role in the paid workforce. Steelworkers felt justified in requiring special treatment at home because they endured long hours and physically exhausting work regimens in mill environments filled with dangerous machinery, intense heat, and deafening noise. Men who worked under these conditions expressed a pride in their willingness to make sacrifices of health and comfort in order to provide adequately for their families.[24]

In exchange for his sacrifices at work, Pauline Pearson's husband expected, and got, the total attention of his wife during their leisure time together. Pauline thought that his request was reasonable within the context of their breadwinner/homemaker marriage: "My husband doesn't want me to run with the girls. Everything we do he wants to do together. If we go to a party he doesn't like it if I get to talking to other people and ignore him." Gail Dobson remembered that her father felt that he was entitled to food service and was sometimes rude about that entitlement: "He insisted that my mother have a meal ready for him whenever he came home, but sometimes after she fixed it he would say he didn't want it." Jean Thompson observed that in the breadwinner/homemaker marriages of the 1950s and 1960s there was always the potential for men to use their breadwinner roles to exercise control over their wives: "My mother never worked. She never did anything or went anywhere. Her husband saw himself as the money-maker; therefore he got what he wanted. He would take naps on the couch and everyone was supposed to be quiet."[25]

In the 1950s it was the automobile that became the centerpiece of the male sphere in breadwinner/homemaker families of steelworkers. Author Ben Herman, a lifelong Dundalk resident, repeated the perspective of the 1950s that, "Cars meant steel, and steel meant us." However, "us" almost always referred to the male breadwinner. Jean Thompson remembered her mother's sense of dependence primarily in terms of her inability to drive: "My mother never drove. It was my father's job to drive us everywhere. My mother could never do anything that required a car." For a steelworker who was earning wages that could finance a nice car, being the sole driver in the family carried with it a great deal of power. Driving his wife to the grocery store or to the shopping center gave him enormous control over when and how well she was able to perform essential shopping responsibilities. Trips to visit friends and relatives could be made only with his approval. Furthermore, her recreation had to be located within walking distance in her neighborhood or else rely on his approval and cooperation. It follows that a woman dependent on her husband for transportation will have formidable constraints on her employment options.[26]

Adult sons and daughters of steelworkers remembered that in the 1950s and 1960s Turner Station and Dundalk homemakers had the advantage of being able to give full attention to child care, housework, shopping, and caring conscientiously for their hardworking husbands. Patricia Cipriani acknowledged that her own mother had been complicit in accepting the constraints placed on her mother as a consequence of the breadwinner/homemaker family structure: "She was one of those women who stayed home and waited on her husband hand and foot. She never worked, never belonged to many groups. She cleaned all day and had a hot meal ready when my dad came home. She felt that women should be at home."[27]

The mill schedule determined all aspects of family life, but most especially the opportunity for wives to work outside the home. Pamela Baneck was matter-of-fact about the extent to which her husband's shift work had kept her from getting a job: "Jimmy doesn't want me to work with his shift work. He likes me to have dinner when he gets home, and I figure he's entitled to that after working ten or twelve hours. Maybe I'll go to work when Jimmy retires." Full-time jobs were discouraged, but even part-time, temporary jobs might be curtailed if they couldn't be adapted to fit into family and shift-work schedules.

FIG. 14 Four young women pose in their best clothes on a Turner Station porch in the 1940s.

Carol Petersen described the way her mother's employment was linked to the schedule at the Point: "My mother would work only during a strike. In 1959 there was a big strike that lasted for an extended period, maybe four or five months, and my mother worked for the telephone company as an evening operator." Jean Thompson remembered that her mother had "felt defeated" when she had to quit the job she had taken during the strike:

> We felt protected because my father worked at Bethlehem Steel and those were very good jobs. We never had any financial problems, but I think my mother suffered, because during a big strike in 1959 my mother was forced to go to work in the personnel department at Hutzler's Department Store. When my father went back to work she had to quit her job and I know she felt defeated. I remember her saying that the job at Hutzler's had given her her own identity.

The "very good jobs" at the Point were a source of protection for steelworkers' families, but they also kept women locked into homemaker roles and out of opportunities to participate in the full-time paid workforce. Wives of steelworkers who did work in the 1950s and 1960s generally did so as a stopgap measure when their husbands were laid off or were out on strike for long periods of time. Part-time jobs as waitresses, sales clerks, and telephone operators were generally the kinds of jobs women took to "fill in" financial gaps that were considered temporary.[28]

When wives in the breadwinner/homemaker families headed by steelworkers got jobs in situations when their husbands were laid off from the Point or went on strike, the circumstances rendered steelworker husbands most vulnerable and least likely to be receptive to their wives going into the paid workforce. Being laid off periodically was a condition of working in the steel industry that steelworkers and their families had to negotiate, despite the fact that it disrupted their lives and could cause considerable mental anguish. An occasional, brief layoff might be used as an extended vacation opportunity if it was brief, but long periods of uncertainty, like the 116-day strike in 1959, could mean financial disaster. Theresa Porsinsky was quick to realize that having her husband, Tom, laid off was both a blessing and a curse: "I went back to work when Tom was laid off, and he took care of the house and kids. It was great in a way, but he was so depressed about not having a job and not bringing home a paycheck that I couldn't really enjoy my job very much."[29]

Cyclical unemployment had an impact on steelworker households that was complicated and that had a different effect on wives and husbands. Women generally understood the dilemma they were in. If their husbands were working at the Point and making a good wage, wives would have considerable financial security and be able to stay at home with their children, but the conditions of shift work would make it extremely difficult for them to participate in any regular educational, occupational, or social activities outside their homes. Having husbands laid off might propel the wives of steelworkers into an expanded social world of work outside their homes. For men, however, being laid off and being stripped of the breadwinner role was an extremely difficult challenge to their masculine identity. A Dundalk teacher reported that, "When there were big lay-offs at the Point I would have to watch out for children being hit, because those were hard times for men who were used to a routine of hard work with a fat pay check at the end of the week." Edna Craig explained this phenomenon as she had experienced it: "Men have gotten used to the control that their job gives them inside their families, and when they lose their job it is not just the job that's gone, their whole world is out of control and it's usually the woman who pays for it. When a man loses his job he's much more likely to hassle the woman he lives with." When a steelworker was laid off from the Point it might compel his wife to get a job, but there was an uneasy sense within the family of a wife displacing the legitimate breadwinner.[30]

When asked about growing up in the 1950s and 1960s, Carol Peterson gave an account of her mother being told to quit a seasonal job, and reflected that the result was a palpable loss, in Carol's eyes, of her mother's autonomy: "One spring my mother wanted to get a part-time job because she wanted to buy some extra things for Easter, so she got a job and she worked three days as a teller. On the third day she missed the bus and got home late and I was sitting at the back door waiting for her. When my father got home he just said, 'That's it. That's it.' The next day she quit her job." Carol's mother was caught in the bind that many post–World War II homemakers experienced. She had the privilege of being supported by a husband who was a breadwinner, which meant that she did not have to be in the paid workforce. But she had also conceded to her husband's demand that she quit a part-time job that had given her some financial power and status in the household.[31]

In many families, husbands and wives developed gender dynamics that encouraged women to stay homebound and dependent. In the breadwinner/homemaker marriages of the 1950s and 1960s, there was always the potential for men to use their economic power to exercise control over their wives. Sometimes the control exercised was motivated by a steelworker's genuine concern that his wife be at home around the clock, because his own schedule made it impossible for him to contribute his time to family concerns.

Jean Thompson remembered that in the breadwinner/homemaker marriages of the 1950s and 1960s, some men used their economic power over their wives in mean-spirited, egotistical ways: "My father-in-law made his wife beg for money. When she started getting a small social security check from the years she worked before marriage, he was angry. He begrudged her that check."[32]

Carol Hartman remembered that her mother's economic dependence was so apparent in their family that it became the focus of mother-daughter tensions: "My mother had no money of her own. It was all the money that my father made at Bethlehem Steel. I can remember getting angry one time and saying to her, 'You don't make any money, you don't put any money in the bank, that's all daddy's money in there.'" Steelworkers' wives who were full-time homemakers responded to the pressures to stay out of the paid workforce by constructing gender identities for themselves that carried considerable status in the 1950s and 1960s in the communities surrounding the Point. Their daughters, however, were beginning to be influenced by the new gender identity emerging in the 1970s, because they had observed the price their mothers had paid for economic dependence. As long as homemaker wives were being supported by the relatively high wages of breadwinner husbands, there were definite constraints on the opportunities that women had to negotiate for more of a partnership in their marriages.[33]

Widows were particularly vulnerable because of the tenacious belief that a steelworker would always be able to protect his family financially, and the fear that a wife who contributed income to the household threatened the family's social status. Because there was general agreement that marriage to a highly paid steelworker meant that a woman did not need to develop any job skills of her own, the widows of steelworkers were especially open to economic difficulties: "When my mother was widowed she had to support herself because my dad's benefits weren't enough. She had never planned on working and had no skills so she went to work in the cafeteria at Dundalk Junior High where she ended up working for thirty years." Like others of her generation, this woman had never anticipated that there might come a time when a breadwinner husband would not support her. By the 1950s, Bethlehem Steel no longer owned a company store or the boardinghouses where earlier generations of widowed women were assured the employment that protected them financially.[34]

In the 1950s and 1960s, a steelworker's wife had money to spend on groceries, on clothing, on home furnishings, and on a family vacation. Measured on a materialistic scale, this ability should have made her content because of the relatively high standard of living afforded her and her family by Bethlehem Steel. But her sense of personal power within her household, within her family, and within her community was restricted by a gender identity and a set of gender relations reinforced by the organization of work at the very steel mill to which she was indebted.

Selma Adams still rankles at the circumstances of her life as the wife of a well-paid steelworker who controlled her personal life by controlling the families' share of his wages:

> It was hard and it was pleasant. It was good being able to stay home and raise my children, but Rudy, he didn't give me any extra money except the food money. It was like being an unpaid maid. If I wanted furniture I had to save up little by little from the food

> money. If I wanted some for myself, I'd have to put away a little each pay until I had enough. That's the way Rudy was, claiming it was his hard-earned money, and his father was the same way with his wife. So we had security with those wages from the mill, but Rudy and his father controlled every penny.

Contemporary Dundalk women like Pamela Baneck express their strong desire for autonomy and power by referring back to their mothers' lives: "I'm as independent as I am because of what I saw happen to my mother. My father was typical of many men in this community. He wouldn't let my mother work, not even a little job in the dime store. And she had to go to him for money for anything she needed, even groceries. I promised myself I would never let that happen to me."

The breadwinner/homemaker marriage that was a product of the prosperity enjoyed by communities adjacent to the Sparrows Point steel mill during the 1950s and 1960s was a double-edged sword. Earlier generations of steelworker wives from the pre–World War II period received universal admiration for the hard work they did, under difficult conditions, taking care of boarders and raising large families without modern conveniences. The economic security that was the great advantage of working at the Point in the 1950s and 1960s afforded the wives of steelworkers a life of less drudgery and more leisure, but those "very good jobs" were also the instrument for keeping them out of the paid workforce and subject to the resentment of husbands and the disapproval of daughters.

During the 1950s and 1960s, male steelworkers were earning time and a half for long hours of overtime, providing well for their families, and certain that America's undisputed dominance in steel production would continue indefinitely. Steelworker families were the beneficiaries of what the USWA had struggled to achieve: the family wage, which enabled one wage earner to support a family in reasonable comfort.

Nonetheless, the daunting constraints of organizing family life around rotating shifts kept steelworker wives out of the paid workforce, and out of other public spheres as well. Steelworkers' wives from breadwinner/homemaker families of the 1950s and 1960s relinquished their autonomy as well as "a working life, a public, social life," in order to fulfill their roles in an economically successful steelworker family. They are remembered by their adult daughters as dependent on their husbands and thwarted in their aspirations. In families where husbands were particularly demanding, daughters and granddaughters describe steelworkers' wives as someone who, sadly, "never worked, never belonged to any groups, never went anywhere;" or who "waited on her husband hand and foot;" or who "didn't drive, we had to depend on my father if we wanted to go anywhere;" or who "didn't have any money of her own."

The extent to which her role as a full-time homemaker was an obstacle to entering the paid workforce is no longer an issue that faces most Turner Station and Dundalk women. The configuration of gender relations in the past, which established male steelworkers as overworked breadwinners while making it necessary that wives of steelworkers remain, throughout their lives, in their roles as homemakers, were undermined in the 1970s and 1980s by the complex economic and social changes impacting this steelmaking community.

5

Women Steelworkers at the Point: Interlopers in a Man's World

I was walking into a man's world. There were calendars with nude women hanging up, and when I started, there weren't even any facilities for women because no women had ever worked in that area of the mill before. . . . It was nothing to be walking along and see some man urinating against the wall. Men have been left alone in that mill for a long time, and some of them [were] barbarians.

—KAREN GRANT, AUGUST 1989

Even before deindustrialization abruptly ended the breadwinner/homemaker family structure among steelworker families, there were a few wives in Turner Station and Dundalk who defied the gender ideology that kept women out of the Sparrows Point steel mill. Beginning as early as World War I, some women attempted to gain access to the male-dominated steel industry, seeking industrial jobs for themselves at the Point. These efforts were limited and repeatedly curtailed, however, by a masculine work culture that only a few women steelworkers have been able to negotiate.

The were some women from Sparrows Point, Turner Station, and Dundalk who were in the full-time paid workforce prior to the 1970s, but for the most part they were unmarried immigrant women, married or single African American women, or widowed or divorced native-born white women.

Some women have tried to make a livelihood at the Point, but until very recently they were always segregated, isolated, or resented by male coworkers. A handful of Turner Station and Dundalk women who worked at the Point have been wives of steelworkers. While interviewing these women I asked how they experienced the day-to-day work environment inside a steel mill. The responses reveal the extent to which a masculine work culture dominated the Sparrows Point steel mill. Through two world wars, the Depression and New Deal, and the unionization of the plant in 1941, steelmaking continued to be "a man's world."[1]

Prior to World War I, Bethlehem Steel began employing women at the Sparrows Point steel mill in one particular job, the inspection of sheets of tin being manufactured for the canning industry. Consistent with the employment patterns for women in the United States throughout the first half of the twentieth century, the cadre of two hundred female employees who worked in that department during each shift were primarily young women who were the daughters of immigrants. This core of women was responsible for quality control, which required "flipping" and inspecting the finished sheets of tin to be sure there were no defects, because the tin would be sold to several large food-processing companies, including Heinz and Campbell's.[2]

In a steel mill where women were excluded from other production jobs, inspecting tin was considered suitable *only* for women because it required a sensitive touch and quick movements of the hand. Furthermore, even though the job was quality control and essential to the production process, the women who did this work were given the nickname of tin floppers. While not consciously demeaning, the term tin flopper removed from the job title any reference to those aspects of the job that required skill and speed.[3]

Sheets of tin were inspected as they were flipped by hand, and finding and discarding sheets that had defects was critical. Tin floppers were hired on a thirty-day trial, and if they hadn't mastered the technique within that period, they were fired. Carol Lane described both the demands and the importance of her job: "When you were flipping—remember it's flying in the air—your eye must be able to see any defects while you turned it over and it flipped down."

Inspecting the sheets of tin was also a job that, like other areas of the steel mill, involved its own dangers: "If you cut yourself on the tin-plate, don't tell anybody, patch it up the best way you knew how, because that was against the rules to cut yourself. We didn't use the company's medical facilities, because then you would have a charge against you that you cut yourself and you were a hazard to this job. We took our own tapes and taped our own cuts. I have many scars to prove it." The use by men at the Point of the term "tin flopper" served to feminize the role of tin inspectors, in part to minimize and belittle the perils of a job within the production area of the steel mill that was physically demanding and yet was done skillfully by women.[4]

Tin flopping was a segregated female niche within a male industry, and between 1925 and the mid-1950s women who did this job were supervised by an imperious forelady who became legendary within the plant and in the community. The descriptions that former tin floppers give of Mrs. Elizabeth Alexander further reveal the extent to which this department of the Sparrows Point plant was feminized by its distinct differences from the prevailing male work culture.[5]

It was Mrs. Alexander's responsibility to oversee the behavior and the productivity of the female tin floppers, and she is described as unique in her success both at rising to the head of a department on her own merits, as well as at controlling a large department of women within a male-dominated workplace: "Mrs. Elizabeth Alexander was head floor lady and I'm sure that no one ever cleaned out her foothills to get her where she was. She had to do that all by herself. She was Hungarian, spoke with an accent, and was tough, really tough. To control 200 women on a single shift you better be tough."

The women inspecting the tin plate were highly regimented, partly as an efficiency measure, but also as a precaution against women being "loose" in a workplace that was constituted as a male social arena. Many retirees of the tin mill remember Mrs. Alexander fondly, but Carol Lane considered her control of the tin floppers dictatorial: "When I went into that sorting room, I must have received the shock of my life because they talk about Germany and Hitler, let me assure you, it was no better down there."[6]

The regimentation of the tin floppers included structured bathroom and coffee breaks. Walking two-by-two, the tin floppers were chaperoned around the plant area like a group of schoolgirls who couldn't be trusted on their own. Matrons supervised the bathroom breaks, another parallel to a strict boarding school intent on maintaining proper female behavior: "Between 6:30 a.m. to 9:15 a.m. you could make one trip to the bathroom, but when you did get to the bathroom they had what they called matrons in there and she could see your feet and if you spent more than twenty seconds in there she'd be pounding on your door telling you to go. Her job was to get you

FIG. 15 An all-female crew of tin floppers working in the tin plate mill in 1938, inspecting the canning plate that was sold to food canning companies like Campbell's and Gerber.

out of there. There was no smoking in there, and no drinking a Coke." The primary purpose of this regimentation was to keep the tin floppers working hard, but the use of matrons was unique to the steel mill's female niche as were the prohibitions against smoking and drinking Cokes on the job.[7]

The other significant purpose for the regimentation was to protect the female tin floppers from flirtatious male steelworkers. Mrs. Alexander assumed responsibility for keeping the women under her supervision physically separate from the men in the tin mill and psychologically suited for proper "lady-like" decorum. She taught her "girls" the appropriate feminine conventions of demure hairstyles and makeup.

In addition, the tin floppers were required to wear uniforms and conform to behavior that connoted respectability: "You didn't chew any gum and you had to wear a blue uniform. You'd better have a crocheted handkerchief in one side of it and it had better be starched. You couldn't come to work sloppy. You had to have make-up on and your hair had to be done. This woman demanded that you come to the tin mill looking great." Mrs. Alexander's concern was that the women who worked under her supervision be seen and treated like ladies.[8]

In contrast to other parts of the tin mill that were infused with a boisterous masculine work culture, tin floppers created a feminized niche in the sorting department that included some of the most traditional female activities. In keeping with the expectations that women contribute to the general social well being, the tin floppers organized fund-raisers at the YMCA to raise money for St. Rita's, the local Catholic church.

The "Kotex Ball" best revealed the anomaly that tin floppers represented in the plant's male work culture. With two hundred women working in the sorting department, Bethlehem Steel was obliged to have sanitary napkins for sale. The proceeds from the sale of Kotex were used to sponsor a dinner dance:

> At the end of the year they gave Mrs. Alexander the money from the Kotex, maybe six or seven hundred dollars and that would be towards our dinner dance and floorshow. You couldn't be in the floorshow unless you worked in the tin mill. We held it at the Belvedere Hotel in Baltimore with something like 600 guests. We had a choreographer and even rented or sewed costumes.

The feminine ritual of a dinner dance clearly was more comparable to a high school prom than to the masculine ritual of stopping at a shop-gate bar, a bar located just outside of Bethlehem Steel property.[9]

Mrs. Alexander's role as both supervisor and headmistress or Mother Superior was evident in the ways in which she provided assistance to poor or immigrant women in her department. She was generous in helping women who were raising children and had financial problems, and she would even buy the required shoes for women who were in dire financial straits.

In activities that would be inconceivable for groups of men working at the Point, Mrs. Alexander arranged cultural outings, including two or three trips every year to New York to see Broadway shows. Bethlehem Steel and the tin floppers themselves considered it completely appropriate that Mrs.

FIG. 16 These retired inspectors and floor ladies from quality control in the tin plate mill had formed a separate, feminine unit—complete with an annual "Kotex Ball"—within the masculine work culture of the Sparrows Point mill. Photo by Jennifer Bishop.

Alexander give the sorting department an element of class and respectability that was the inverse of the male work culture elsewhere in the plant.[10]

Even Rosie the Riveter found it difficult to gain entrée in significant numbers at the Sparrows Point mill. Large numbers of women living in Sparrows Point, Dundalk, and Turner Station went to work during World War II, replacing men who went into the service. Few of those women worked at Sparrows Point in production jobs.[11]

Compared to 35,000 women who entered Baltimore's shipbuilding, aircraft, and machinery industries, only 700 women took production jobs at the steel mill during World War II. Mary Anderson, the director of the Women's Bureau of the U.S. Department of Labor, forwarded the following report to Secretary of Labor Frances Perkins explaining why so few women had gone into the steel industry:

> Steelmaking traditionally has been men's business. Steelmaking is a heavy and dirty business and women workers have been taboo.

> . . . [I]ntense heat and heavy equipment are necessary in processing. These marked characteristics of the industry and inherent hazards have tended naturally to shut out women, with their lesser strength and endurance. . . . [W]omen have constituted only a fraction of 1 percent of the employees in the steel industry. Steelmen—both managers and workers—generally did not welcome the advent of women into their mills and feared that women would not be able to do a full job and would be a disrupting element and liability. . . . Also, there was a deeply rooted prejudice and tradition against women workers in the steel mills similar to that which prevails in the mining industry.

The "deeply rooted prejudice and tradition against women workers in the steel mills" meant that many of the women who worked at the Point during World War II endured considerable harassment. Only a very tough woman who was willing to be a maverick could last in a workplace where they were considered "a disrupting element and liability."[12]

Janet Lawrence was part of the worker migration to Baltimore from West Virginia during World War II to seek employment in war-related industries. After working as a tin flopper, she moved up to become an expediter, a job that paid more money but where there were no other women:

> I particularly was working with a fellah, his name was Jimmy and he was the Campbell Soup shipper. The only thing they allowed me to do was total the tickets, and whether I was right or wrong he would throw them in the air and then he would sit at a calculator and re-total them to make me look bad. One day Jimmy said to me, "What the hell you working for, isn't your husband working? Why didn't you marry a man who has a job?"

It took a stubborn woman with some imagination to figure out ways to get around the ironclad grip that white men had on the day-to-day control of the plant and on the selection of cronies for advancement at the Point. Black men faced this obstacle as well, because "the company didn't support women or blacks. Blacks had menial jobs and women had menial jobs. You had to know somebody to get a promotion or even a job there."[13]

Janet Lawrence did get to know somebody who was a supervisor and also from her hometown. He could not control the sabotaging behavior of the men Janet worked with, but he was able to do her a favor that enabled her to outsmart her male coworkers:

> My supervisor moved me to export where we had two wild men. They would start early in the morning screaming and yelling that they couldn't possibly get their work done. One fellah was an alcoholic; he was always in the bathroom nipping on a bottle. The other one wouldn't talk so I had another struggle, which was to learn export. They wouldn't tell me what the tariffs were, they wouldn't tell me anything I needed to know to do the job. When the supervisor asked how he could help, I said, "Change my hours. Bring me in two hours later than them so I can have two hours with their paperwork after they leave." After that I worked two hours later and researched the paperwork and did a lot of homework. That was the only way that I learned export shipping, because those two fellahs were never going to show me!

Refusing to train a new coworker was cited repeatedly as a strategy for keeping women and blacks out of better-paying jobs.[14]

There also was a set of behaviors at the Point that were so pervasive and so masculine that it made it difficult for women to fit in. One of these behaviors was language. The language of a workplace serves several functions, one of which is to establish a climate. Newcomers must adapt to that climate or leave. The climate of rowdy, vulgar language at the Point was intimidating to many women who simply did not know how to interact with a communication style foreign to their own. Vulgar language could also be used to intimidate women, just as steelworkers often used it in attempts to bully one another. Reflecting on her work as an expediter at Sparrows Point during World War II, Janet tells a story of her repeated failures in a workplace controlled by men, until she learned how to use the potent weapon of vulgar language:

> An expediter has to go all the way back into the mills and check the product as it's coming through to make sure it's there on time for shipment. The offices were full of men, no women, and here comes this one, Janet Lawrence, who is going to be an expediter. Well, I missed every shipment that I was supposed to expedite, because they would tie it up at the washers, they would tie it up at the plating line, they would tie it up at the packaging and there wasn't

> a thing I could do about it. At the end of the week, you had a chart with how many customers you made deliveries to this week. Consequently, I had zero. All my materials were tied up.

As had been true since Rufus Wood complained about favoritism toward Catholics in June 1889, cliques ran different departments of the mill, and the men who were expediters were solidly opposed to a woman joining the ranks of higher paid steelworkers. Janet came to the realization that one of the ways in which she did not fit was in her way of communicating: "The language they used wasn't good morning or how are you. The language they used was f—— you, you son-of-a-bitch, you bastard. All day long. I wasn't raised that way, so I didn't buy it."[15]

Because vulgar language was the vehicle that many steelworkers used to distinguish themselves as tough, and to prevent a woman from taking a "man's job," Janet faced a choice of giving up and taking a lower-paying job, or fighting on ground that was unfamiliar to her: "I went down into the black plate department and walked into Mr. Huffman's office and said, 'Mr. Huffman, may I see you a few moments?' And he said, 'What the f——, you think I have all the time in the world for you? I got a million f——' jobs goin' here. Get the f—— outta my office.'" With the moment of decision facing her, Jane initially admitted defeat. She walked halfway across the yard in retreat, and then stopped:

> I turned myself around and went back to black plate and didn't open the door, but kicked it in. I kicked that door in and he jumped up from his chair, his eyes bigger than saucers, and I said, Listen mother-f——, these orders are coming through or I'm going to blow your f——' head off." And I walked out. Don't you know my orders came through and I had only a couple of misses that week? That was the door I opened and finally I was getting all my orders and I became a good expediter, the first woman expediter. My superintendent said, "I'm very proud of you . . . how did you do it?" I said, "By using their language." So I found out that I had to meet them on their level; it was no longer, "Good morning, Joe," but, "How the f—— are ya, Joe."

Janet learned the lesson that all women who work in steel mills and other male-dominated industrial workplaces must learn. Steelworkers go to church, help their neighbors, teach their children to be polite, and work for charitable causes. But inside the steel mill it is a man's world and the use of rough language is unrestrained. Janet was able to adjust to that; many women cannot.[16]

Although Janet was one of the 700 women who successfully held jobs at the Sparrows Point steel mill during World War II, the end of the war marked an abrupt end to women's employment in production at the Point, except for the crew of women working in quality control in the tin mill. In 1945, like other wartime employment for women in industry, steelmaking jobs at the Point once again became the purview of men.[17]

The World War II–era campaign by the SWOC likewise failed to challenge gender segregation in the steel industry. Although the SWOC was militant in its battle for an improved hour and wage system at Sparrows Point, it did not question the conventions of gender that had existed at the plant since the 1880s. Women made up only about two percent of the membership of the SWOC, although it is important to note that large numbers of wives of steelworkers supported the union effort. Between 1941 and the 1970s, female membership in the USWA remained at about two percent, reflecting the absence of women in the steel mill.[18]

Through the 1970s, the union's newspaper, *Steel Labor,* acknowledged its female readers with a monthly dress pattern and set of recipes, and it acknowledged the gender consciousness of its male readers with a monthly bikini-clad pinup who met the two necessary qualifications: being full-breasted and a member of one of the unions that served traditionally female occupations, such as airline stewardesses or waitresses. Only a handful of articles referred obliquely to the feminist issues of the decade, and none of them discussed the desirability of hiring more women on production jobs in steel.[19]

Throughout the post–World War II era, the organization of work at the Point was based on a race and gender hierarchy that went unchallenged by the USWA. The USWA was dedicated to improving wages, benefits, and working conditions, but for several reasons it declined to address issues of gender inequities, and it was able to achieve only occasional victories in gaining equity for black steelworkers. The local and national culture was imbued with an assumption that well-paying, skilled industrial jobs were the prerogative of white men.[20]

During the post-war period, the USWA was committed to the idea of the family wage, a concept that represented a wage scale sufficient to enable a man to provide for his family as the single wage earner. Encouraging women to work outside the home was in logical opposition to the concept of the family wage, so to champion the hiring of women at Sparrows Point would contradict one of the primary goals of the USWA. So it was not the USWA but, rather, the Consent Decree of 1974, won by the organizational and legal efforts of black male steelworkers, that opened the doors at Sparrows Point to women steelworkers.[21]

Only a few of the women in Turner Station and Dundalk chose to leave homemaking roles to take advantage of the access to a range of jobs at the Point. For a woman to take a job in production at Sparrows Point required the willingness to do a job that the larger society viewed as an inappropriate feminine role, and for this reason many wives of steelworkers eschewed work in the mill for more traditional jobs as nurses, teachers, secretaries, bank tellers, and retail clerks. Women who had grown up in a steelmaking community were well aware of the resistance of male steelworkers to having women at the Point.[22]

Women who were mavericks and willing to enter nontraditional roles, or women who were on their own and in need of industrial pay for their families, were the ones most likely to seek jobs at the Point in the 1970s. Some of these women were exceptionally large and strong, while others were petite and dainty. Many of the jobs inside a steel mill, though, including operating a crane, require skill and attentiveness rather than inordinate strength. Karen Grant, one of the ten women steelworkers interviewed, was the daughter of a steelworker and the wife of a steelworker.[23]

Grant, a woman who began working at Bethlehem Steel at the age of eighteen in 1976, is a good example of a young Dundalk wife who was able to "tough it out" at the Point. Karen had lived her whole life in Dundalk on a block that was close enough to the mill that she grew up listening to the slag being dumped into the water on summer nights. During her childhood every man on her block except two worked at Bethlehem Steel, but Karen never expected that she would do production work at the Point. Her twenty-six-year career as a steelworker began when her younger brother graduated from high school:

> He asked me if I would take him down there and I said, "Sure, and while I'm down there I'll put in an application too." My husband laughed. He thought that was hysterical. "You? You work at Bethlehem Steel? You wouldn't last five minutes down there. You couldn't even pass the test." As a joke, I put in an application. It was at the time when they had a quota and had to hire a certain percentage of women. When I got a letter asking me to come down and take the

> test my brother, who didn't get a letter, was mad. He said, "That's no fair. It's just because you're a girl."

Karen took the test and passed it, went for the interview and was hired and given a work assignment:

> Someone from the employment office called me and told me that I got the job and that I would be starting in the steelmaking labor department, which is the number four open hearth, which is bull-work; that is real hard work. My husband said, "Oh God, Oh my God. On the steel side? You'll never be able to handle it." My husband worked on the tin side, which was not such extreme work.

Despite her husband's doubts, Karen did establish herself as a skilled and hardworking steelworker.[24]

Because women had not performed manual labor in the production area since the World War II period, the Consent Decree resulted in an influx of women who had not planned to have careers as steelworkers. Karen was a young married woman who hoped that she and her husband could earn enough at Bethlehem Steel to buy a house. Ramona Smith was a thirty-year-old black woman from Turner Station who was on her own, the sole support for her family, and desperate for a job that paid the kind of wages that would cover her bills and provide health insurance. Ramona described her motivation for seeking a job at the Point:

> Prior to working at Bethlehem Steel I was working as an assistant manager at Sambo's Restaurant, and I was working the graveyard shift. My daughter, Holly, was going on two, my former husband was completely out of the picture, and I was trying to get off of welfare. I heard about jobs at the Point through the unemployment office. Bethlehem Steel was close to home, it was more money, and the benefits were good, including BlueCross/BlueShield. I was called down to Bethlehem Steel to be tested. From there I was hired as a bricklayer's helper in the labor department in March of 1979.

Ramona's desire to get off welfare was the most compelling factor pushing her to work at the Point.[25]

Like many men and women before her, the good pay at Bethlehem Steel in comparison to other employers impressed Karen:

> The day I got my first paycheck they must have thought I had gone crazy, because I had never seen that kind of money in my life. They handed me that check and I looked at it and screamed: "Oh my God. This is mine? All this is mine?" It was $350 clear, and that was the hook right there. I was going to be there for the duration. I knew I would be making good money but I had no idea that I would be making that kind of money. Back in 1976 that was a lot of money. That was the real thing that hooked me in completely because I was still undecided, but when I got that paycheck, that was it.

Union wages at the Point that had previously been totally inaccessible to women were partly at the root of the resentment men expressed as they watched women get jobs that they themselves were denied. White men had lost their cherished monopoly on the wages that could be earned in the steel industry.[26]

Both Karen and Ramona were emphatic about the fear that they initially felt. The men I interviewed minimized their fears. For men, facing stoically the hardships of working in steel was an essential component of the ethos of manliness. Ramona was candid and articulate about the fear she experienced:

> We got the metal-toed shoes, a hard hat, a briefing on safety tips, and our assignments. I was scheduled to work any shift that they gave me. We got a tour of Bethlehem Steel and the area where we would be working, in the open hearth. I was scared to death. It was a dangerous situation. You might be working down in the hole and have a forklift coming in and a train coming in behind that with scrap metal. The furnace next to you would be lit up to 1200 or 1500 degrees. You're talking about something so hot that it's almost like lava coming out of a volcano. I went home and prayed, asking God to please help me stick this one out.

In the 1970s there were deep resentments on the part of many male steelworkers toward the women who were hired under federal edict to work at the Point. The jobs most frequently given to women, including crane operator, had served as an informal niche for older steelworkers who had put in a lot of years of hard work and were thought to deserve lighter assignments in the years before retirement. The fact that the government imposed quotas was compounded by the fact that Bethlehem Steel, like other steel companies, had fought hard against the Consent Decree rather than conscientiously preparing for female workers. Karen Grant started working in the number four open hearth with a group of about 15 other women at a time when the company was so ill-prepared for female workers that they didn't even have bathroom facilities for women. The women were given makeshift facilities in what had been a foremen's locker room.[27]

Both women and men confirm that there was active antagonism from male steelworkers to the presence of women who began doing production work at the Point in the 1970s, and even after working in the mill for years, many women who were interviewed in the 1980s acknowledged their status as outsiders: "There is a lot of rejection toward women. There is such a masculine factor in this job. Some guys will say right out that they don't want to work with me." Karen described her early experiences working in the production of raw steel as a struggle to overcome the resistance and taunting of her male coworkers:

> You had to take these real long rods and hook them up to these cables that are three inches around and throw the cables up on to the back of a huge truck with tires as big as me. You have to have some upper body strength to get that cable up there. I kept trying to throw those cables up on the truck, but I didn't have the strength and they kept flopping off. I looked over and there was a group of guys watching me and laughing. They thought it was the funniest thing. Not one of them would get up and help me out. So, being the determined nut that I am I kept trying and trying until after about five times I'm getting real disgusted and I turned to the men and said, "You think it's funny, don't you." I went over and got a ladder and put it against the back of the truck, dragged the cable all the way up the ladder and onto the truck. When I was done I set the ladder next to the men and said, "Funny, huh?"

Only women who had the courage to stand up for themselves were able to persevere against pervasive hostility.[28]

In the 1970s, women were considered interlopers at the Point. Most white male steelworkers viewed the Consent Decree as a preposterous anomaly perpetrated by federal courts that represented liberal, civil rights, and feminist interests. One woman discussed her forceful attempts to disabuse her male coworkers of their assumptions that all women steelworkers had a political agenda:

> When we were first hired they took every woman down there as ultra feminists who burned their bras. I explained to one guy: "I'm not down here to prove any point. I couldn't burn my bra if I wanted to, I couldn't live without it." They just naturally assume that if you're doing what they consider to be a man's job, then you're trying to be a man, or you're a man-hater. I explain to them, "Look, I bake cookies. I have two little children. I'm no different than your wife." If I'm dressed up and going out, no one would know that I'm a steelworker, but some of the men expect you to have a cigar hanging out of your mouth.

Women steelworkers are aware that the work culture within the mill has significant historical foundations and that they are moving into an environment that has been formed by decades of men working without the presence of women.[29]

The ways in which women are treated by their male coworkers varies considerably, and one woman speculated that about a third of the men made efforts to help women get established in the mill, about a third were relatively indifferent, and another third were actively antagonistic. Accusations launched at Ramona Smith from hostile coworkers were direct and fierce:

> My first assignment was in the number four open hearth, working as a slagger. A lot of those men really weren't used to working with women and a lot of guys resented us; they really didn't want us around. There were also a lot of guys who were happy for us to be there. You could usually tell the difference. Some of them would come right out and tell you. I was sitting in the lunch room one day and this one guy said, "How old did you say your baby was?" and I said, "Eight months old." He said, "What the hell you doing down here? Ain't you got no sense? Why don't you get home and take care of your kid where you belong?" I was surprised and hurt so I just said, "I don't really know you." Today I would tell him, "You have no right to tell me where I belong, and my children get just as hungry as yours." If I had my way I probably would be glad to sit home and eat bonbons and watch TV, but that never has worked out too well for me.

Male steelworkers who were wedded to the breadwinner/homemaker family structure experienced women production workers in the mill as a threat to what they were accustomed to as a proper social arrangement.[30]

For other men the issue was one of distrust that women could or would do their share of the work. Some men refused to speak to women in their work units, regardless of how well those women performed or how hard they worked:

> One man on our crew was friendly with everyone else but he was always cold and distant with me and the other girls. So one day I just came out and asked him, "You don't like me, do you?" He said, "No, it's not that I don't like you, it's just that I don't think you should be here because I don't think you can do the work. I think you make it a lot harder for all the other men." So I said, "Well, I have just as much right to be down here as you do and I *can* do the work. What I can't do in brawn I make up for in brains." He didn't say anything, but he still didn't talk to any of the women.

Most women who work at the Point do their best to work hard and be competent members of their work crews. There are accounts, however, of women who came into the mill wearing elaborate hairdos and long painted fingernails, expecting to use feminine wiles to get male coworkers to carry their share of the work. Their manipulative behavior caused resentment among some men.[31]

Often it was the wives of steelworkers who resisted most vehemently to women working at the Point. Few wives will admit their fears that their own husbands might begin socializing with female coworkers and date or have affairs with them, but that undercurrent of jealousy does exist. When the mill was an all-male workplace, wives could send their husbands off to a job that excluded the temptation of female allure. Many women who have worked in steel are happily married or make it a policy, as Karen Grant did, not to socialize with coworkers:

> The worst thing you can do is flaunt the fact that you are a woman. If you're wearing the tight sweater you're inviting trouble. The women who get all upset when they break a fingernail are not going to get the respect of the men. And when men ask you to go with them for a drink after work, it's usually a sexual thing. They don't want to just sit and talk with you. I always say no because

> getting involved with somebody on the job is a big mistake and leads to trouble with the other guys.[32]

Like Janet Lawrence during World War II, women who began working at the Point were harassed with rough language. The women who succeeded at the mill have had to find creative ways to control or temper language they find offensive:

> The men test you with language. They talk filthy if you let them. I've had this happen to me many times. I'll be sitting in the lunchroom and there'll be one particular guy who thinks he is getting to you. So he'll go on about what he did to his girlfriend or wife last night, and then he'll look over at you to see if it's affecting you. I usually ignore it, or you might have another guy in the room who'll say, "Hey, don't talk like that, you got ladies sittin' here." If I have to address it I look them dead in the eye and say, "How you talk is not my business. I'm not your Mommy, just don't direct it at me." Some people tell me I shouldn't let them talk like that around me, but, look, it's not my lunchroom and I'm not going to lower myself to answer them back with the same filthy language.

Along with sexually aggressive language, pictures of nude women are a symbol that male steelworkers use to mark the steel mill as a man's world where women can be freely denigrated as sexual objects. Gwen Reynolds, a white woman who began working at the Point in the 1970s, was unnerved by the pinups she encountered at the Point, but she refused to let them intimidate her:

> I walk into offices and I'm shocked by the posters of nude women they have hanging up. If I don't go in they'll have accomplished what they want to accomplish, which is to keep me out. There is a culture down there that says, "I'm a real man." It's a badge. A lot of guys can't read *Playboy Magazine* at home, but he's got his locker stuffed with them.

Women steelworkers view the rough talk and the pinups as part of a cult of masculinity inside the mill, a kind of group behavior the purpose of which is to impress other men: "The men are out to impress each other. They talk different when women aren't around. They need to prove their masculinity.

You'll find yourself talking about gardening with a man, but if another man walks up they catch themselves and start talking about 'let's go for a beer.'" In the 1970s and 1980s, the masculine work culture at the mill was a major obstacle to male steelworkers crossing boundaries of gender and race to interact authentically with all of their coworkers.[33]

Ramona Smith linked the sexual harassment to men drinking at the Point, and she also felt most vulnerable when she was working the night shift: "The worst experience was some of the men drinking. If I was working on a midnight turn some of the men propositioned me for sexual favors. You tell them to get away and they're walking behind you, it's at night and it's pitch black out there and they could grab you around any corner." Because it sometimes was a union official that was involved in the harassment, Ramona found it fruitless to pursue the grievance process that was intended to protect women as well as men. She did, however, find that she had allies among male steelworkers:

> For you to file a grievance or anything, you never really got the action that you expected. I even had to report one of the shop stewards because he kept pulling his zipper down and tucking his shirt in right in front of me. It wasn't that I could see anything it was just that it was as if I was going to see something. Anywhere I turned this man was doing it. I was working with two black men that day, and they kept saying to me, "Rae, you shouldn't have to take that." I'm embarrassed as all get out. I finally said, "You're right." I asked the foreman, "Would you please do me a favor, and ask so-and-so to please keep his pants up for once today. I don't need to see this."

This kind of fierce antagonism to the presence of women at the Point perpetuated women steelworkers' outsider status. It discourages most women from considering working in a steel mill in the first place, and the harassment they experienced from men was responsible for more than a few women quitting jobs in steel.[34]

In the last decade, strict enforcement of sexual harassment policies and laws has diminished overt harassment. Male steelworkers remember the introduction of women in larger numbers to the mill in the 1970s as a period of bitter discord. Some men who now work amicably with women steelworkers can vividly recall their own participation in the harassment of female coworkers. In 2002, both men and women report that female steelworkers are generally accepted and often liked and admired as members of the paid workforce at the Point.[35]

Nonetheless, very few women have gained entrée to industrial work and pay at the Point. Beginning with the tin floppers and Rosie the Riveter, women had small or temporary niches in steelmaking. In the 1970s, when women won legal entitlement to compete for production jobs at Bethlehem Steel, jobs in steel were declining rapidly, and today only a few hundred women work in production at the Point. The women who remain have certain qualities as survivors, one of which is their ability to accommodate brusque male behavior. Gloria Robinson, who works as a trainer for new employees, was sure that she "could never work out there with all that heat and dirt and all that roughness." The hazardous working conditions at Bethlehem Steel, combined with the aggressiveness of social relationships between male steelworkers, gives the Sparrows Point steel mill a gendered workplace environment that is palpably masculine and unappealing to most women. Beginning with the inspectors in the tin mill, small numbers of women have worked at the Sparrows Point steel mill since before World War I, initially in a segregated department, and today in production jobs throughout the plant. These women are best viewed as pioneers and trailblazers who have worked hard to be successful in what continues to be a man's world.[36]

6

Deindustrialization at Sparrows Point: Disappearance of the Breadwinner/Homemaker Family, 1970–2000

We try to instill that they shouldn't look at a layoff as a tragedy, and the better majority have picked up the pieces. For a lot of guys it was a very difficult adjustment. Guys didn't want their wives to work. At least they can take solace in the fact that they weren't alone. You have to grab hold of the idea that it is out of necessity, that it is something you have to do, that there is no shame in it. But for lots of guys that's a hard thing to do.

—DOUG INGLES, JULY 1989

Deindustrialization transformed the communities surrounding the Sparrows Point steel mill. Beginning in the 1970s, the prosperity that steelmaking had brought to Turner Station and Dundalk and to the breadwinner/homemaker family structure that depended on that prosperity began to unravel. In contrast to the massive closings of mills around cities like Pittsburgh and Buffalo, deindustrialization assumed a different face at Sparrows Point, which continues to be a highly productive mill despite a dramatically reduced paid workforce. Also, unlike rust belt areas where steel was the only employer, Baltimore's sprawling metropolitan area offers other kinds of employment to both women and men. As jobs for men at the Sparrows Point steel mill declined over the past three decades, Turner Station and Dundalk women began entering the paid workforce in large numbers, altering significantly the gender consciousness and gender relations in the community.[1]

After nearly a century of being able to rely on Bethlehem Steel, Turner Station and Dundalk families were confronted with the reality that steelmaking would no longer provide for the employment needs of their communities. Competition from foreign steel and from producers of materials like aluminum and plastic pushed the Sparrows Point mill to modernize in ways that required fewer employees. Turner Station and Dundalk families began looking elsewhere for jobs to support their households. After two decades of steel prosperity that supported the breadwinner/homemaker family, wives of steelworkers resumed the economic roles that their grandmothers had once fulfilled. These modern wives became partners in the wage-earning efforts of their households. While Sparrows Point wives at the beginning of the twentieth century had managed boarding and rooming businesses within their homes, at the end of the century wives in the communities surrounding the Sparrows Point steel mill left their homes and went into the paid workforce in response to deindustrialization.

Since 1980 it has become common for the Sparrows Point steel mill and the surrounding industrial communities to be viewed through a postindustrial lens that assumes that industrial work has vanished from America. When I tell people from other parts of Maryland that I am researching the Sparrows Point steel mill, the most common response is, "You mean Bethlehem Steel is still open?" or "Does anybody still work there?" Many Americans who work and live far from industrial centers believe that steelmaking is a thing of the past in the United States, and that steelmaking communities are poverty-stricken backwaters.

Much of what has changed about the Turner Station and Dundalk community can, in fact, be viewed as decline. Before 1980, steelmaking was the economic lifeline of this area and, in fact, Bethlehem Steel ranked as the single largest employer in the state of Maryland. But by the 1980s and 1990s, the daily newspaper carried regular reports of cutbacks, layoffs, and falling profits. Membership in the USWA locals 2609 and 2610, both of which represent workers from the Sparrows Point mill, declined from more than 12,000 members in 1970 to 4,300 members in 1993. In March 2001, Bethlehem Steel announced its support for a major industry consolidation after nine steelmakers filed for bankruptcy protection in 1999 and 2000.[2]

According to the 2000 census, this steelmaking community has lost population as well. In many areas, homes for purchase are being replaced by rental properties, and jobs outside the community coupled with the trend toward suburban sprawl have drawn residents of all ages to the rapidly growing outlying areas of Belair, Joppatowne, and Timonium. Turner Station and Dundalk have lost prominent shopping establishments, because the East Point and White Marsh malls northeast of the Sparrows Point area draw customers away.

In the 1960s Sparrows Point, Dundalk, and Turner Station families assumed that their sons would leave high school and enter the steel mill, where decent wages would support a family, a family home, and a homemaker wife. Today, youth unemployment is one of the community's most vexing problems. The average income for the community has declined, and the number of families who live on poverty-level incomes has risen. There are increasing numbers of subsidized housing units and low-rent apartment buildings springing up in a community that historically has prided itself in home ownership. The *Baltimore Sun* ran a front-page article on a group of homeless men who had set up camp in one of Dundalk's parks. And most Turner Station and Dundalk families know someone in their circle of acquaintances who has filed for bankruptcy.[3]

Yet despite all of the warning signs of economic and social distress, the decline of employment at the Point has not spelled disaster for the community. Dundalk and Turner Station continue their heritage of vigorous residential neighborhoods where men and women now pursue employment opportunities outside of the Sparrows Point mill. Community activism continues to rally citizens around political, educational, and housing issues. Both the Dundalk-Patapsco Neck Historical Society and Museum and the Turner Station Heritage Society maintain a strong commitment to the heritage of the area. The Fourth of July Parade in Dundalk rivals any in the state of Maryland. The hard work ethic continues to be the prevailing value in the community, but by the end of the twentieth and at the beginning of the twenty-first centuries, women had joined men from Dundalk and Turner Station in bringing that ethic into the paid workforce.[4]

The story of deindustrialization is now a familiar one in American life, and social problems including youth unemployment, drugs, and crime are now a challenge in both Dundalk and Turner Station, just as they are in other post-industrial communities. However, alongside the issues created by the loss of thousands of industrial jobs with union wages has come a reconfiguration of family structure and gender relations re-establishing the economic participation of women in households in Turner Station and Dundalk. The convergence of steel's decline with a popular consciousness more sympathetic to women created the conditions necessary for women and men to renegotiate the set of gender relations that had characterized this community in the 1950s and 1960s, replacing the male breadwinner with a partnership of husband and wife, both of whom make essential contributions to the household economy. In an extraordinarily short span of time, no more than thirty years, the sole provider that had been a creation of post–World War II industrial prosperity vanished in families in Dundalk and Turner Station because all of the conditions that created the male breadwinner have disappeared from these communities.[5]

As late as 1970, Dundalk and Turner Station could be definitively identified as communities of steelworkers. The 1970 census indicated that for all of Baltimore County a total of 5,838 men were employed in the primary metal trades. Out of that total, 5,198 lived in Dundalk and Turner Station. Beginning with the 1980 census, employment statistics were listed for the whole of Baltimore County but not for specific communities within the county. These census figures show a steady decline in the manufacturing sector between 1970 and 2000, when manufacturing jobs in Baltimore County decreased from 66,500 to 35,400.[6]

This process of deindustrialization has been mirrored by the Sparrows Point plant, which has seen a steady decline in number of employees that has shaken the entire metropolitan region. In its 1977–78 Directory of Major Employers, the Greater Baltimore Committee listed 20,000 employees for Bethlehem Steel's Sparrows Point Plant. The 1981 Directory shows a slight decline to 19,500. In the 1983–84 directory, however, the number of employees dropped to 15,000. From 1988 on, no listing whatsoever appears for the steel mill, even though the Sparrows Point plant continued operating. In the mid-1990s, Sparrows Point and Burns Harbor were Bethlehem Steel's most profitable plants, but the profits were due to modernization and automation that had resulted in drastic eliminations of steelworkers. The *Baltimore Sun* reported 6,000 Sparrows Point employees in 1993; 5,200 employees in 1997; and 3,500 employees at the beginning of 2002. In the two decades between 1983 and 2002, employment at the Point mill has declined by 80 percent.[7]

The structure of the entire economy of Baltimore County has shifted in ways that are characteristic of deindustrialization. Between 1970 and 2000, manufacturing jobs in Baltimore County decreased by almost 50 percent. During this same thirty-year period, jobs in the retail trade increased from 36,800 to 83,400; jobs in finance, insurance, and real estate rose from 10,200 to 41,700; and jobs in the services area soared from 35,800 to

160,500. Well-paid jobs in manufacturing that could support a breadwinner/homemaker family structure were replaced with jobs that were not unionized and that for the most part had a wage scale that required two incomes to support a family comfortably.[8]

During this same thirty-year period, households in which women contributed to the family income were replacing the breadwinner/homemaker family. Between 1970 and 2000, male participation in the labor force in Baltimore County decreased from 83.1 percent to 75.7 percent. Female participation in the labor force in Baltimore County during this same thirty-year period increased from 42.9 percent to 62.8 percent.[9]

The decline of households with a single breadwinner is also documented in census data. In 1970, the majority of households in the Baltimore metropolitan region—57 percent—had a single male breadwinner. By 1980 that percentage had slipped to 42 percent, and during the 1980s it dropped to about a third. By 1990 only 36 percent of households in the Baltimore metropolitan area had a single breadwinner. Over the same thirty years, wives in the Baltimore metropolitan area were entering the paid workforce in large numbers. In 1970 wives in only 43 percent of households were working; by 1979 this figure had risen to 58 percent. In 1990, 64 percent of households in the Baltimore metropolitan area had two people in the workforce.[10]

Because of changes in the display of census data over the last thirty years, it is not possible to provide specific numbers for the communities surrounding the Sparrows Point steel mill. By reaching back to 1950, however, we know that the overwhelming majority of families in the Turner Station and Dundalk communities sent a male breadwinner to one of the 30,000 jobs at Sparrows Point, while only 22 percent of women in these communities were in the paid workforce. Today the 3,500 jobs at the Point can no longer sustain breadwinner/homemaker families for the communities surrounding the Sparrows Point plant.[11]

The concept of the "male breadwinner"—the hardworking husband who supports his family comfortably on his earnings from the Point or from other industrial employment—has virtually vanished from discussions of family life in Turner Station and Dundalk. There are a number of families where the husband *is* the only person in the labor force, but this arrangement has become the exception rather than the rule. Even more indicative of a change in community mores is the fact that young women in their late teens and early twenties imagine for themselves futures that are based on the assumption that they will be in the paid workforce whether or not they marry. The dwindling numbers of jobs that remain at the Point no longer make it possible for a breadwinner to be a reliable source of family income throughout his lifetime.[12]

The loss of thousands of steelmaking jobs between 1970 and 2002 created a crisis in this community that was faced reluctantly and painfully. Donna Prince, whose family was active in the USWA, described the mindset that accompanied the era of abundant steel jobs at Bethlehem Steel:

> The Point created a way of life that people have a hard time recovering from. When I was growing up that was the place to work because it meant lots of money, new cars, boats and Ocean City vacations. Working at the Point was a good way of life. It provided well for families. But it instilled an attitude that someone would take care of you. It was Fat City. We created this culture that was so easy and so expected. Anything after that wouldn't be as good or as well paid and it would be a struggle because we'd have to learn something new.

In ways similar to other single-industry communities, the neighborhoods belonging to Sparrows Point steelworker families faced the decline of their economic underpinning.[13]

In interviews with men from the communities surrounding Sparrows Point, laid-off steelworkers describe the ways in which deindustrialization has forced them to reconceptualize the economic strategies necessary to support their families. Even men who currently provide all of their families' income are reluctant to talk about themselves as breadwinners because they are aware that the day may come when economic changes will make it impossible for them to fulfill that role. For men whose fathers and grandfathers have been steelworkers and breadwinners, the process of relinquishing the role of provider has often been difficult.[14]

Those individuals and groups—particularly the USWA—that have decried the loss of men's jobs, have accurately predicted that the loss of well-paid industrial jobs would hurt many Dundalk and Turner Station families. The Displaced Workers' Project, sponsored jointly by the company and the union, provides support and retraining to men whose careers in steel were abruptly ended. The results are mixed.[15]

Men who have found other permanent, full-time jobs have almost always had to take a substantial cut in pay; the impact of this has depended upon the length of time they had worked at the Point. Men in their twenties were more able to retrain or adjust their lifestyles. Men in their fifties who had spent their entire lives assuming that, like their fathers and uncles, they would always be making good union wages, have had more difficulty. Many men have suffered from depression, desperation, and physical illnesses. But after facing the realization that their employment at the Point was finished, most went on to other jobs, adjusting to the change as best they could.

Those who found jobs at lower pay with a long commute are considered by their peers to be the success stories, in comparison to the men who continue to "lie on the couch waiting for the company to call them back." Describing his coworkers who could not give up the illusion that jobs might once again be plentiful at the Point, Bob Shaw commented: "Some guys go from being the breadwinner to being the loaf. In most cases they are satisfied to do little jobs. They don't want to be tied down with a nickel and dime job after the money they made at the Point." Unable to make the transition from steelmaking to another full-time occupation, these men are the real casualties of downsizing at the Point.

However, others who have gone through the shock of losing their jobs in steel have ultimately been glad to be forced out of steelmaking and into other kinds of jobs. Bethlehem Steel had the lure of good wages, but many families of aging steelworkers are deeply resentful of the disabilities and early deaths that the Point's retirees have suffered.

Even union activists admit that steelworkers and their families paid a big price for their high wages. The USWA was able to demand decent wages and other benefits for steelworkers, but the exhausting system of the swing shift and the physically and psychologically enervating conditions inside the mill were beyond the power of the union to ameliorate. As a result, men, while well paid, spent their lives in the mill facing health hazards, debilitating working conditions, and little chance for upward job mobility.[16]

From the men and women of Turner Station and Dundalk who are working in other kinds of employment has emerged more criticism of steelmaking jobs. Their new jobs have helped them realize that going to work at the Point was frequently an easy, seductive choice. Doug Ingles, who spent twenty years at the Point, found that, once he began working there, it was hard to leave:

> A lot of people get stuck. They're there, it's a paycheck, and they know it's coming, and they just get used to it and they're stuck there. They don't take a chance and they don't venture out to do anything else because that's what they know how to do. They get used to it, and unless they get laid off for an extended amount of

> time and they get scared and they need money, they don't do anything else.

Doug had to go through the anxiety of not being able to provide for his family, but he had this to say about being laid off: "When you're working a job like the mill you tend to become satisfied and you don't realize that you can blossom. My head got completely turned around by the money I was making at the Point. The lay-off was an opportunity for me. I've been able to do things that I never thought I could." Reflecting on his new skills and the new opportunities for developing his potential in areas of human relations, talent that had been completely stifled in the competitive, harsh environment at the Point, Doug reflected: "Sometimes I think losing my job at the Point was a blessing in disguise. Now I have a chance to get a job in a place where people are nice to each other." He found that a job where he could develop new skills was more important to him than a hefty paycheck at a job that was grinding him down.

Repeatedly, men reflected that they were ultimately grateful to have been able to make a transition to jobs that provided lower pay but more challenges to learn new skills. Larry Sayles was in his thirties when he lost his job at Bethlehem Steel. Like others, Larry expressed gratitude for the chance to grow in his work: "Looking back, it was almost like a blessing for me when I got laid off from there, because it made me go on to other things. I ended up working at a bakery as a first class maintenance mechanic, and that job just felt sweet to me after Bethlehem Steel. Now I'm at a radio station in public relations, and it's a whole different world. I've been able to grow since I left the Point."

New jobs that did not include shift work also made possible an entirely new and much more manageable home life. Previously, Larry Sayles had felt chronic frustration with his job in steel:

> For couples that wanted to build relationships that were healthy, shift work was always a challenge. He comes home at 11:00 P.M. with energy to burn. She has had a bruising day at home, wants to sleep. They try to share some responsibilities, but when he's working three-to-eleven, most of the work of ferrying kids around, and doing the errands, falls on her. And the school activities are strictly her responsibility because he's at work during the soccer games and the school plays.

When he began working a day job with another industrial employer, both he and his wife reported that the opportunity for a more egalitarian arrangement of family responsibilities compensated for the income lost from juggling family life and shift work.[17]

Husbands who have been laid off from the Point and have made successful changes in employment and lifestyle invariably talk about wives who are encouraging and enterprising. These men are married to women who can cope with adversity, wives who come from generations of hardworking women. Faced with a radically changed economy and no longer living in a social world that encourages their reliance on a breadwinner husband, women from Turner Station and Dundalk have joined the paid workforce in large numbers.[18]

The transition has not been easy. In interviews, these women describe the fears that accompany the loss of their husbands' jobs. They also describe an array of personal, cultural, and educational obstacles that had functioned to keep previous generations of women out of the paid workforce.

Women who have been pushed into the paid workforce by layoffs affecting their husbands are adamant that reliance on a breadwinner is no longer an option for their families. Angela Giordano described the experience of losing the financial security her family had enjoyed for twenty years: "[if my husband] wanted to work overtime he could, if he wanted to work weekends, he could." Until the 1980s Angela assumed that, just like her mother and her aunts, she would spend her life as a homemaker because, "It was never a problem that I had to have a job." Angela's initial reaction was shock: "I just said, 'Oh, my God.' I knew I was going to have to go to work full-time. All my close friends are going through this." For Angela and the women who are her friends, the realization that she must make an essential contribution to the household income was initially shocking.

The women and men that I interviewed expressed the mix of emotions that descended on Dundalk and Turner Station families in the midst of a major economic and personal transition: concern for husbands who faced depression and low self-esteem as the result of job insecurity; fear on the part of wives who felt poorly prepared to enter the paid workforce; and, ultimately, pride on the part of successfully employed women and acceptance on the part of those men who have adjusted to the new household economy.[19]

When I asked Donna and Walt Czerwinsky about their transition to a two-wage household, their responses were complicated. They had relied on Walt's job at Bethlehem Steel for seventeen years before he got laid off. Walt described the sense of humiliation that he, as a person who had lost a high-paying job, felt when his wife experienced the excitement of successfully entering the paid workforce. Donna described Walt's reaction to her prospect of full-time work while he resorted to a series of low-paying, temporary jobs:

> When I suggested that I go to work, he had a really funny feeling. He got very depressed. Going in the unemployment line and looking for other jobs, whether he admitted it or not, was very, very depressing. One job he took with a guy across the street, he had to ride to Washington in the back of a truck in the summertime. The job was painting, and you didn't have to be a painter. If you could hold a paintbrush and just put your hand up and down you were hired. He hated it, but it was that or nothing.

Donna's own experience of going into the paid workforce was difficult as well, but, unlike her husband, she was moving into, rather than being pushed out of, the position of wage-earner: "I really enjoyed going back to work. It gave me a real good feeling. I used to feel like I was just a housewife, and that I didn't know how to do anything."

For both husbands and wives this was more than an adjustment in lifestyle. Having wives go into the full-time paid workforce involved a complete rearrangement of gender roles and gender relations that forced each member of the couple to think differently about the assumptions on which their marriage was based. For most couples this adjustment was difficult; for some it was unachievable. In every case new gender identities were established within the family.[20]

A man who was stripped of his breadwinner role as a result of industrial layoffs had two adjustments to make. He had to accept his new lower status of employment, and he also had to accept the fact that he was reliant on a wife whose income was now necessary for the family's financial survival. Jacqueline Wallace recalled her husband's difficulty in reconciling himself to a radically different household wage-earning arrangement:

> When Pete was laid off it was hard for him to go from making twelve dollars or fourteen dollars an hour down to minimum wage. He had a wife who was making more than he was. He had a lot to diminish his self-esteem. There was a time when I got a very good job and he was acting real peculiar. He indicated that he didn't

> know me anymore. I had to tell him that I married him to be my husband, not my father.

The conflict within the Wallace family represents one of the complex realities of deindustrialization, a readjustment that is repeated in one steelworker family after another. Pete, having spent his entire adult life as a provider with precise skills who works inordinately hard for a good wage, is now stripped of his status and pay as a steelworker and his role as breadwinner. Jacqueline, in the tradition of steelworkers' wives throughout the history of Sparrows Point, is prepared to make the necessary adjustments in a time of crisis. She is willing to go into the paid workforce, and now the previous obstacles to her participation in the full-time paid workforce have been eliminated: her husband no longer makes an income that enables his wife to stay out of the paid workforce, and he is no longer working the rotating shifts that made it impossible for his wife to maintain a regular job. When his wife says, "I had to tell him that I married him to be my husband, not my father," she is also participating in a renegotiation of gender roles by reframing the marital relationship from one that is patriarchal to one that is more egalitarian.[21]

In 1900, George S. Webster looked at Sparrows Point through the masculine lens of nineteenth-century social commentary when he entered the names of Sparrows Point wives into the manuscript census but indicated that they had no occupations. Historically, women in this community have always worked hard. Between 1887 and World War I, Sparrows Point wives were cast into the role of businesswomen managing small boarding establishments within their homes when alternative forms of housing for single men were not available. After World War I and into the 1950s, women in Sparrows Point, Dundalk, and Turner Station continued, albeit in fewer numbers, to take in boarders. Widows were given jobs in the company's cafeterias or cooked and cleaned for the local priest. Women in the African American community often supplemented income from boarders by providing domestic services in the homes of white women who were able to pay for ironing, cooking, or cleaning. During World War II black and white women from these communities worked in the steel, aircraft, and electronics industries.[22]

The years between 1950 and 1970 were a short-lived anomaly, in many ways, because during this period the role of the homemaker who was not bringing income into the household became a widespread phenomenon. This was especially true in Dundalk, but even in Turner Station, where

FIG. 17 Three generations of women and children in a Dundalk family in 1958. Courtesy Family of George and June McNeill.

employment for married women continued to be more common, there were black steelworkers whose income supported a breadwinner/homemaker household. Daughters of homemakers married to breadwinners in these communities describe their parents' relationships as paternalistic, in part because community values privileged the male breadwinner role and stigmatized married women working outside the home.[23]

The life trajectory for married women in this steelmaking community between 1950 and 1970 was predictable. As young girls they completed educations designed to fit them for their domestic responsibilities. They married soon after high school, raised children and provided domestic services that accommodated the erratic schedules and arduous work that their husbands did at the Point. They made essential contributions as mostly unpaid homemakers to the functioning of a highly productive paid workforce of steelworkers.

Daughters of the women who were raising children in Turner Station and Dundalk between 1950 and 1970 describe their mothers as women of remarkable achievements. All of these achievements are in the tradition of women's domestic support for male steelworkers. Daughters describe their homemaker mothers as devoted to childrearing, housekeeping, and providing support for overworked husbands. In a community with a long and remembered history of frugality, women recall with admiration the ways in which their mothers budgeted the household income. The achievements of post–World War II homemakers married to steelworkers consisted of services rendered to various members of the family and rooted in the tradition of self-sacrifice on behalf of others.[24]

Just how pervasive the standard of the breadwinner/homemaker family was can be discerned when women discuss how completely their own mothers disapproved of married women going into the full-time paid workforce. Some women, like Yolanda Martin, report that it was their own mothers who offered the most formidable resistance to the idea of employment outside the home: "After I got married and started having children I didn't work. When I went back it was just to see if I could get out in the business world again. My youngest son was ten and I started out working two or three days a week. I fibbed to my mother. Even though I was fifty-three years old she disapproved of me working, so I told her that it was just temporary." In this atmosphere of community sensibilities and generational ties that functioned to keep women homebound, going out into the world of work became a process that required a revision of deeply entrenched community mores. Going to work meant abandoning what had been defined as a wife's rightful place.[25]

During the years between 1950 and 1970, when the Sparrows Point steel mill was a source of seemingly endless employment possibilities for men, educational institutions also discouraged young women from considering careers in the paid workforce. Local high school programs and career counseling prepared young men for the reality of abundant jobs in the steel mill; programs and counseling for young women readied them for marriage. The system of tracking in the high schools in southeastern Baltimore County accommodated the needs of the steel industry, something that local educational institutions have always done. The implicit message to young women, however, was one that devalued their skills and abilities. Willa Fisher, who completed her training as a nurse when she was in her forties, remembered bitterly the advice she got from her high school counselor:

> In the eleventh grade the guidance counselor called me down to her office. She didn't even ask me to come in and sit down, she

> just let me stand out in the hall, and she said, "You're not planning to go to college, are you?" And I said "No." And she said, "Good, because you'd never make it. You should get married, have babies, that's all you'll be able to do." I believed it for so long. When I look back I see that if you didn't come up to the highest standards, you were pushed aside, you were put on the back burner.

Even when young women actively wanted careers, counselors discouraged them. Carol Peterson, who had clear career goals and was highly motivated to go to college, was urged to abandon her aspirations: "I was hell bent on going to college because I was going to be a gym teacher, but my eleventh grade counselor talked me out of it because she said there were too many teachers and by the time I graduate I won't be able to find a job. Sometimes I'm sorry, because I would have been a good gym teacher."

Women in Turner Station and Dundalk repeatedly give accounts of authority figures telling them during the 1950s and 1960s to lower their expectations. The messages that parents, teachers, and high school counselors gave to young women essentially encouraged them to assume that their skills were most suited to homemaking and that marriage to steelworkers was the most reasonable choice for their future security, given the economic dominance of steel. Women were never urged to seriously assess the marketability of the skills they had.[26]

For the most part, the women I interviewed who had married steelworkers in the 1950s and 1960s had not chafed against internalizing marriage and motherhood as a cherished ideal. Amy Brown, who now excels at her career in job placement, remembered her goals when she was in her twenties: "As far as I was concerned my life only began when I got married. That was what I was waiting for, to be a wife, to be a mother. Then everything would be o.k." Amy had no career plans when she married, and she assumed that her life would proceed much as her mother's had. Nonetheless, being pushed into the paid workforce by unforeseen economic and social changes led her to opportunities that she relished and that she knew her mother did not have.

The detached researcher may view the entry of women into the labor force as part of a widespread social trend. To wives of steelworkers who married young with the expectation—absorbed from families, schools, and communities—of being supported by well-paid breadwinners, entering the full-time paid workforce was a critical life change. It led these women to a much-revised perception of themselves as participants in their families, as members of the paid workforce, and as citizens in their communities.[27]

For most Turner Station and Dundalk women, entering the full-time paid workforce after 1970 was not a rejection of marriage, family life, or household responsibilities. These women do, however, make a distinction between the home-centered responsibilities of housework and child care, and the more public worlds in which they participate now that they earn a wage, make a substantial contribution to their households, and develop broader social networks. Willa Martin made a clear distinction between working women and housewives:

> The working makes the difference. And today there are still plenty of women who do not work outside the home. I have two such women who are my children's in-laws. They are night and day to me, these women. They're both younger than I am, but I have almost nothing in common with them. When we get together for the little ones' birthdays, they would have everything in common with my mother because they were all housewives. And, heck, I cook and do all those things, too, but I just have a whole different set of things that I care about because I'm out. Working has a big effect on you.

Belonging to the public world of the paid workforce created a social network.[28]

In interviews with wives from Turner Station and Dundalk who entered the paid workforce in their 30s, 40s, and 50s, I asked questions about the benefits of having full-time jobs and about the advantages and disadvantages of the homemaker roles that they had been forced to give up. I was interested in how they thought about the process of balancing their work responsibilities and their family obligations. What were the stresses related to their jobs that contributed not to a growth of skills, but to personal and family conflict? What had they been able to accomplish that was of value to themselves as members of the full-time paid workforce? Was there a personal satisfaction in earning a paycheck? Did they have a sense of being important in the world beyond the family?

Gail Dobson talked about her growing knowledge of computers and the sense of personal power she derived from helping her male boss with his computer problems:

> I have a lot of knowledge in different areas because I have worked a number of different jobs. We've changed computer systems at

> my present job, and I feel competent because I have a computer background. The other day my boss came to me and said, 'I need some expert advice,' and he asked me to help him figure out what he had done wrong. That made me feel good because I could show him how to correct his program.

For many wives who have gone into the paid workforce as the steel industry declined, job experiences are a source of personal power that goes beyond the financial benefit of the paychecks. They say that the rewards inherent in demonstrating competence and skill in the workplace are as important as the economic rewards. To have knowledge and expertise that is valued by coworkers and supervisors is a source of both job satisfaction and an enhanced sense of self-esteem, something that distinguishes these contemporary working women from their mothers.[29]

Many women, like Gail, found jobs in companies where the work culture is far more egalitarian than the hierarchical culture of a mammoth steel mill. In a non-industrial and team-oriented work environment, the self-confidence that a female employee can develop and her willingness to assert her authority begin to blur formal hierarchies. When asked what made her feel most competent, Angela Giordano described several occasions when she relied on her own judgment in a confrontation with someone who technically was in a position superior to her own: "Gretchen wanted to take over the desk of an employee who was on leave and I looked her in the eye and said, 'I don't think that's a very good idea.' When she challenged me by saying that she'd discuss it with my supervisor, I stood firm: 'We don't have to ask Jim, I know what I'm talking about.'" Taking a stand on job-related issues and earning the respect of coworkers are negotiated differently in an office or a commercial enterprise than they are in a steel mill. Angela based much of her sense of accomplishment on her expertise at coordinating complicated projects and knowing the entire workplace well enough to take initiative. She commanded respect from coworkers in higher status positions and even from the general manager: "We work as colleagues, and we serve as a sounding board for one another. I can question his decisions and make reasonable demands about my work situation. He listens to me because he knows that if I complain it's for a good reason." For Angela, the fact that she was highly respected by her colleagues and her boss was of utmost importance. Being valued for her competence, having some control over her work situation, and participating in a cooperative work environment were all sources of her sense of self-esteem.[30]

For many women in Turner Station and Dundalk, employment has meant not only leaving their homes but also leaving their communities. In the 1970s cars in Dundalk sported bumper stickers that announced the imperative to "Live, Work, Shop Dundalk." The slogan was a symbol of the pride that Dundalk residents had in a cohesive community that was able to provide for all the needs of its citizens, and also able to keep out the disorder of the larger world. Historically, one of the advantages of living in Dundalk or Sparrows Point or Turner Station was that there was little reason to leave the area, because every necessity could be found within the community, often within walking distance. Yet beginning in the 1970s, large numbers of women from southeast Baltimore County began commuting into downtown Baltimore to work as secretaries and receptionists in the city's law firms, doctors' offices, and government agencies. Donna Czerwinsky described her determination to get a job downtown despite her husband's disapproval:

> When I had an opportunity to get a job, my husband said, "Don't go downtown. That's crazy. You don't have a car or anything. How are you going to get there?" I said, "I'll take a bus." I was determined. It would take a long time on the bus, but I still wanted to do it. My husband wanted me just to stick to the Dundalk area. That's the way he was. I said, "No. I'll try Dundalk, but if I can't get it, I'll take whatever I can get." It was downtown, and he sort of frowned about it, but he didn't try to stop me.

Ruth Shriver made an explicit connection between overcoming her fears about a job downtown and her growing sense of self-confidence: "When I started my new job downtown I was driving through bad sections of the city to get to work. I was feeling very self-reliant. I was going to city nightspots on my own after work. I felt really good about myself. I even lost forty pounds."

In fact, many women from this steelmaking community talk about the process of going into the full-time paid workforce as one of the most important transitions in their lives. They acknowledge the stresses that accompany the attempt to work full-time along with raising families, but they see their jobs not as an oppressive burden, but as an opportunity to be "important to more than just my family." Donna Czerwinsky expressed her sympathy for her husband's loss of self-esteem when he was laid off from Bethlehem Steel, but she went on to talk about her own enthusiasm at having the chance to go into the paid workforce: "The first chance I got to go out to work, I jumped at that chance. It just gives me a good feeling that I'm capable of doing something. I get praises and I've gotten promotions without asking for them. It gives me a very good feeling, to hear someone say that I'm a good worker. They evaluate us once a year, and I've always gotten a good evaluation, and it just makes me feel very, very good." One woman remembered her joy at getting her first full-time job: "It was almost an unbelievable thing that I had gotten this job. It made me feel wonderful. I was so proud of myself." For many adult women in this steelmaking community, a full-time job serves as proof of her competence, but it is also a symbol of newfound economic power that neither she nor any other women in her family had ever experienced before.[31]

Angela Giordano saw going into the paid workforce as an opportunity to contribute and to be important in a world that goes beyond her role in her family: "When I didn't work and my husband supported the family, I felt that the housework should be my responsibility. I enjoyed staying home and I didn't mind being the one to do the housework, for a time. But I also enjoy being important out in the world. Having a job makes me feel as though I am important to more than just my family, and I like that feeling."

In addition to more economic power and independence, more equality in their marriages, and a greater sense of competence, participation in the paid workforce has been a release from the isolation that had characterized previous generations of steelworkers' wives. One woman asserted that her job was beneficial because it pushed her beyond the "small world" of her parents and grandparents who were "almost like carbon copies of one another" because they all lived in the same neighborhood and stayed within a small group of neighbors and family members. Established in the workplace and living more independent lives, women from the Sparrows Point area view themselves as a group that has improved its standing in the community. They look around their community and see women everywhere who "know that they can do so much more than the kitchen work."[32]

However, this perspective on the changes that have occurred for women in the neighborhoods surrounding the Point is at sharp variance with the approach underlying the community's simultaneous discussion about the loss of steelmaking and other manufacturing jobs. Both perspectives are based on real experiences, but one focuses primarily on the experiences of men who have lost high-paying jobs in steel, while the other focuses primarily on the experiences of women who have established a public social identity and a sense of self-esteem as newcomers to the paid workforce.[33]

The decrease in employment at Sparrows Point has presented challenges to the community surrounding the plant, and the reconfiguration of family economic strategies has resulted in both benefits and losses for women. For those women who have entered the paid workforce in order to contribute to the economic survival or prosperity of their households, the changes in family dynamics have often meant an increase in personal competence and an escape from the constraints experienced by the homebound wives of earlier generations of steelworkers. It has also meant renegotiating family systems where two breadwinners are juggling the tasks of full-time work, child care, shopping, cooking, cleaning, and the maintenance of cars and homes.

7

Renegotiating Families with Two Breadwinners: Partnership and Divorce

We arranged our work schedules so that he was home when I was at work, so that one of us was always with the kids. When I left for work, whatever I didn't finish I left for him. He's good with cooking and helping the kids at bath time. He has no patience with homework, so I try to have that finished before I leave. He does all the laundry, and taking the kids to recreation is something we do together.

—PAMELA BANECK, SEPTEMBER 1988

Contemporary women from the communities surrounding the Sparrows Point mill describe their own marriages as fundamentally different from the marriages of the homemakers of the 1950s and 1960s. Typically, contemporary wives did marry right after high school, and assumed that their marriages would follow the pattern of their parents' generation, with a steelworker husband supporting a homemaker wife and their children on a single industrial wage. The post–World War II era of middle-class breadwinner/homemaker families relishing a good life at the height of America's steel productivity turned out, however, to be a short-lived period of prosperity for the families of Sparrows Point's steelworkers.

Deindustrialization forced contemporary wives into the full-time paid workforce as the number of jobs at the Point plummeted from a peak of

30,000 in the mid-1950s to slightly more than 3,000 in 2002. Contemporary wives in the communities adjacent to the Point must bring income into the household not as a stopgap measure, but in order for their families to pay mortgages and car loans. In this respect, wives in the communities surrounding the Point have lives and marriages that are strikingly different from that of the post–World War II generation, and more closely parallel the economic roles of the wives in the company town of Sparrows Point in 1900 who were managing small boardinghouses to earn necessary family income.

What adult daughters remember of marriages in the 1950s and 1960s is the overriding importance that was placed on men as the breadwinners, while family life and gender roles in steelworker households of that era were defined primarily around the concept of men being the "head of household," a term that appeared in the U.S. census for every decade until 1970. The vernacular for "head of household" was the ubiquitous boast from steelworkers in the 1950s and 1960s, "My wife doesn't have to work," a statement that implied a set of gender relations that included the understanding that wives of steelworkers could work in times of crisis, such as a long strike or layoff. Wives could also work occasionally to contribute supplemental amounts to the household for holidays or special purchases. However, if a man was not the sole breadwinner, or if a man's wife was supporting the family financially, it suggested failure on his part to be a successful provider.[1]

Men's work schedules and earning power gave them entitlements within the family to which wives acceded in the 1950s and 1960s. However, adult daughters remember that a family system organized around the male providers' boast that "my wife doesn't have to work" was often detrimental to their mothers' sense of autonomy and self-worth. The entitlements that accompanied well-paid jobs in steel included the assumption that a man who works hard has a right to expect a hot meal at the end of his workday, whenever that might be, with the consequences that the shift work schedules of steelworker husbands determined when the family ate and kept women out of the paid workforce even after their children were grown. Work schedules at the Point also relegated child care to wives. Husbands generally were absent from their children's lives, and as chronically tired steelworkers they claimed the prerogative to demand that wives keep the kids quiet to accommodate the mill's shift schedules. Wives maintained the household budget, and buying a house was almost always a joint decision, although it was common for the wife's name to be omitted from the mortgage and deed. Similarly, the utility bill was almost always in the breadwinner's name. Because it conformed to the logic of post–World War II gender relations that men "should be in the driver's seat," steelworker husbands purchased, maintained, and drove the family car.[2]

In the 1950s and 1960s, a wife's primary responsibilities included housework and caring for children. She was responsible for less household drudgery than her grandmother's generation, but was given the responsibility for acquiring the abundance of new consumer goods that became available after World War II. Being the family's shopper was a double-edged sword, however. Steelworker families were proud that they had an income that allowed them access to the new products, which in the 1950s became necessary for the family's status. Gloria Stone, who never bought an automatic dryer because she enjoyed hanging her clothes outside to dry, was asked constantly, "What's the matter, can't you afford a dryer?" But shopping could be devalued as a frivolous activity, because along with essentials like laundry equipment came inessentials like makeup and guest towels for the bathroom, and women's roles as consumers did not gain the same respect as being in the paid workforce.[3]

When the wives from the communities surrounding the Point went into the full-time paid workforce in the 1970s, they began receiving something that few wives of Sparrows Point steelworkers had ever had before—a paycheck. Wives who took in boarders in the company town of Sparrows Point had earned money that was necessary for the household budget, but it came in the form of cash from the boarders' pay envelopes and was put into the family's collective household funds. A paycheck with a woman's name on it is a symbol of individual financial power, and only with deindustrialization did paychecks become widely available to women in the communities surrounding the Point.

Pamela Baneck described her elation the first time she got paid: "My first paycheck was only $40 but I felt so big, I was making plans to buy drapes and furniture and re-do my rooms. I felt so good about earning that money, it could have been $400 instead of $40."

In the course of my interviews I asked women and men in the communities adjacent to the Sparrows Point steel mill how deindustrialization has forced contemporary families to renegotiate gender roles in the following areas: the change from breadwinner households to dual-income households; ownership and control of the family car; responsibility for meals, housework, and child care; and the purchase of major consumer items—cars, air conditioners, even homes—by women instead of men.[4]

Prior to World War II, steelworkers' wives had little discretionary income and few consumer items to choose from. Steelworkers' wives in the 1950s and 1960s had responsibility for spending a significant portion of the family's income, but could be criticized for purchases that the breadwinner considered extravagant. Wives like Carol Peterson who went into the full-time paid workforce after 1970 reported that they used their roles in the paid workforce as the basis for claims to partnerships within their marriages and control over their spending: "In my household it's a partnership. It's very different than the household I grew up in. My father worked at Bethlehem Steel, and he controlled everything. In my household, I work every day, I work just as hard as my husband does. I have to get up and get out of the house every morning, and if I feel like buying something, I feel like I have the right to spend the money I've earned." Carol described her paychecks as an endorsement to choose purchases that her husband could not criticize as frivolous: "I feel really good about getting a paycheck. This is my money and I like controlling it. I know I can charge on my American Express and I can pay the bill when it comes. I have a car phone now and I know I can afford the expense."

When I asked men and women how husbands who were former steelworkers responded to losing their breadwinner status, many couples remembered that when a wife entered the paid workforce her paycheck could be a source of disputes. Patricia Cipriani recalled how her excitement at receiving her first paycheck clashed with her husband's difficulty in accepting the new arrangement of a two-paycheck household:

> I was so happy to be bringing home a paycheck; I ran up to the credit union and opened a separate account in my name only. When I told my husband I was really excited. This was something of my own. He didn't say anything, but he started acting funny. Just getting sort of sarcastic and making comments like, "Well since you have your own checking account that nobody else can use, I guess you can pay these bills."

Patricia was typical of dozens of wives who expressed their enthusiasm about getting a paycheck and being able to open a bank account in their own names. The reaction of Patricia's husband was characteristic of heads of households in the process of losing their breadwinner status. Over a period of several years, Patricia's husband was reassured by the fact that so many wives in the community were going into the paid workforce, while

Patricia's initial excitement was tempered by the combination of satisfactions and burdens that came with a full-time job.[5]

In some families former breadwinners who saw their wives moving into the full-time paid workforce assumed control over the household finances to maintain a measure of power. Fran Smith described an elaborate drama that took place each time her husband paid the bills: "My husband always made paying the bills sound like such a scary thing. He would sit at the table and pay the car insurance and say (she imitates a deep, gravelly voice), 'This is such a hard job.' Then he'd pay the mortgage and wipe his forehead and say (she lowers her voice again), 'I don't know how I'm going to do this.'" Ultimately, Fran recognized that her husband's ritualized warnings about the hazards of paying bills and maintaining a household budget were his way of resisting the reality that he would now have to share the breadwinner role.[6]

So how did having women in full-time jobs change the gender dynamics in households previously headed by steelworkers who were breadwinners? Wives in the communities near the Point recount that going into the full-time paid workforce did not, by itself, transform their marriages into egalitarian partnerships. When wives began to gain economic leverage and access to the public arena through their participation in the paid workforce, a process of renegotiation began that challenged each element of the gender relations and gender expectations of the previous generation. How much husbands resisted changes in gender roles and how much wives persisted in maneuvering for more independence varies among families.[7]

Both men and women describe the difficulty of renegotiating gender roles. In Sue Giordano's family the ownership of the family home became the focus of the negotiations between husband and wife over making their marriage more egalitarian. For Sue, having her name on the deed to the family home symbolized more equality. For her husband it symbolized a loss of his economic power in the family and meant giving up control over the household that he wished to head: "My husband refused to put my name on the deed to our house. I would bring it up and he would say he couldn't do it, it was too expensive. I did some research on my own and found out it would only cost $75.00." The story did not end there, because Sue was determined to change an aspect of her marriage that had come to represent her subordination to her husband. She confessed that she consciously used nagging (defined brilliantly by Webster's dictionary as "persistent small assaults") to get her name on the deed to the family home: "I made that man miserable. Not a day went by when I didn't bring up the issue until finally he got tired of hearing it." Her victory on this seemingly small issue was significant in two ways. She had engaged her husband in negotiations over a financial issue that her mother's generation would never have questioned. Secondly, she designed and implemented a deliberate and ultimately successful plan of action to achieve a change in the power dynamics of her marriage.[8]

Making decisions about major consumer purchases is an arena in which Turner Station and Dundalk women who are divorced or unmarried flex their muscles in an effort to achieve more self-determination as consumers—not of breakfast cereals and laundry detergents, but of major, "manly" items like cars and furnaces and air conditioners. This is a community where a lot of men earn their living installing furnaces and air conditioners, repairing washing machines, or remodeling kitchens. The purchase of major consumer items are unequivocally categorized within the male sphere of expertise, and in the 1950s and 1960s women did not cross that line: "A big purchase was never made by a woman alone when I was growing up. That was something men did and women never even concerned themselves with those kinds of things. If you were widowed, you called your son-in-law to buy an air-conditioner for you if you needed one." Claiming competence in the area of buying major pieces of equipment that in the past were purchased exclusively by men was a source of pride for Edna Craig: "I was very proud of myself when I bought a new furnace and new air conditioners for my house. Men take women seriously now because we have money to spend and can buy things on our own."[9]

Buying a house of one's own is probably the ultimate step in negotiating the arena of financial skill and self-determination. Fran Smith bought her own home in the 1980s, when she was both divorced and middle-aged. The process of assuming a major financial responsibility transformed her sense of self: "My house is small, but to me it is the most beautiful house in the world. I never thought that I could buy a house on my own, but I did, and I know that I can take care of it, too. Now I don't think there is anything I couldn't do." Like other women who have bought their own homes, Fran's house represents economic self-sufficiency and individual autonomy.

In a community where adult men grew up with mothers who were dependent on their husbands for transportation, a wife who wants to learn to drive can appear to her husband to be a frightening deviation from the kind of marriage in which he expected superiority. A woman who knows how to drive has the potential to dramatically decrease her dependence on her husband and dramatically increase her autonomy. A wife who is learning to drive can create a crisis if her husband believes that the stability of their marriage relies on his wife being dependent on him for money, transportation, and expertise with repairs and business negotiations. Joan Zagorski concluded that only their involvement with their church had saved her and her husband from divorce: "When I learned to drive and got my driver's license my husband was so insecure. He was sure that our marriage was doomed because I was being so independent."[10]

Jean Thompson remembered that until she was in her late thirties she depended on her husband to drive her wherever she needed to go, making it necessary for her to "beg him to take me places." For Jean, learning to drive became the first milestone in a long effort to be more independent, and she started out taking lessons secretly because of her husband's intense objections to her having the mobility to go places on her own: "I didn't learn to drive until 1979. I saved my own money to pay for the lessons and I made appointments with the driving instructor when my husband wasn't home because he resented everything I did that showed independence."

Learning to drive was more than a new skill acquired as a matter of course. It was closely linked to Jean's determination, however tentative at this point, to achieve more personal autonomy: "I was scared to death to drive, sure that I wouldn't be able to learn. But I had a Helen Reddy tape with 'I Am Woman' on it, and before the driving instructor came to my house I would play that tape again and again to overcome my fears." Women who learn to drive also change the gender dynamics within their marriages, and Joyce bolstered her determination to drive with messages that broadcast a radically different gender consciousness than the one with which she had grown up and married.[11]

In the breadwinner/homemaker families of the 1950s and 1960s, an automobile represented a man's pride in his paycheck and his role as head of the family. Buying and driving a car was so definitely a male prerogative that even a divorced woman would often let fathers or sons or next-door neighbors negotiate the purchase for them. Jean Thompson talked about negotiating the purchase of a new car for the first time as a significant mark of her independence and a major accomplishment in self-sufficiency:

> I was really proud of the first car I bought for myself. I knew I needed a car but I kept putting it off. Then one Friday afternoon I left work and said, "This is it." I changed my clothes. I thought to myself, "I have to look a certain role." I went to Norris Ford and told the salesman, "This is what I can afford." I left there within an

> hour with a fancy Pinto, and drove straight to my mother's house to show her.

She was justifiably proud of an accomplishment that her mother's generation of women would never have considered possible. The purchase of her own automobile was such a significant act that it requires deliberate preparation, including a suitable costume, and it is a transition that, even five years later, she remembered as a sign of independence. Joyce also cited buying a new car and maintaining the repairs on it as a gauge of her competence: "Buying a new car was a terrific thing for someone who hadn't been driving for that many years. I took it by myself to be repaired and fixed. This was good for my sons to see that I could do these things on my own."[12]

Descriptions by Turner Station and Dundalk women of their conscious efforts to make their relationships with their husbands more egalitarian are filled with anecdotes about the deliberate strategies they continue to use in order to accomplish relatively small changes in their marriages. When I began this project I was looking for ways in which the feminist movement had manifested itself in a steelworkers' community. I found in the mid-1980s what persists to this day—a taboo on the word "feminism" that is not unlike the taboo a century earlier against women entering the steel mill. When Barbara Orlitzky suggested to her Bible study group at a Dundalk church that she was interested in the spiritual contributions of women, the deacon retorted with alarm, "That sounds like a *feminist* point-of-view." When she urged her community organization to involve more women, the weapon used to attempt to silence her was, "What are you, some kind of feminist?"

Wives in this community are practical, sturdy, and smart about the need to be flexible in order to accomplish goals. I have rarely heard a discussion about feminism in interviews, in community meetings, or in the classroom. Nonetheless, women actively seek goals designed to gain more independence and autonomy for themselves, and to get husbands to relinquish some control or assume more responsibility.[13]

Pamela Baneck is aware that she is engaged in a process of changing the assumptions upon which her marriage had been based. She sees herself engaged in a process of renegotiating her marriage and reframing her husband's consciousness about gender relations. She values her ability to be creative about the changes she makes in her marriage, and will often design her strategies so that they are acceptable alternatives to a direct confrontation. As an example, she recalled how she had circumvented the delicate issue of visiting with her husband's mother every Saturday night: "I didn't like doing this. I wanted some time away from my husband and kids, but I knew if I told him that his feelings would be hurt. So I volunteered to run the Bingo at church, and he thought that was important so *he* took the kids to his mother's himself. [She smiled broadly.] I was free for four hours every Saturday night!" Pamela is aware of how radically she is challenging the customary dynamics of the breadwinner/homemaker family, and she understands the need for patience: "Most women are too impatient. My husband is the kind you don't pull surprises on. I learned that early in my marriage. I pave the way for myself. When I want to do something I start talking to him way ahead of time. I don't wait until I'm ready to do something and then spring it on him." Time and again women recounted how carefully and thoughtfully they had planned their attempts to renegotiate gender relations within their marriages.[14]

During interviews with wives in the Sparrows Point area, I asked whether family-based friendship networks were still the most common social experience of contemporary women. The responses I got attest to the fact that support networks today are based in the workplace or at church or community meetings, and they are oriented toward encouraging more personal independence and more egalitarian marriages. Having a support network that provides encouragement or solace is important to Turner Station and Dundalk women who are in the process of reconfiguring the gender dynamics in their families, and some women have groups of friends who have been together for decades: "We have been meeting together and socializing together for more than 25 years now, and we have been through a lot together. No matter what the crisis—a pregnant daughter, an unemployed son, an alcoholic husband, or any other hardship—you know that this group of women will listen to you and be sympathetic to your situation. Sometimes that's all it takes to get through it." Support networks are built at work, in the neighborhood, within the family, or even at church.[15]

Some women participate in groups that socialize after work in much the same way that their husbands used to drink together after a shift. While these groups are casual excursions to eateries around the metropolitan area, they also symbolize a claim to certain freedoms that were denied to the homebound steelworker wives of earlier generations. Donna Czerwinsky explained the ideology behind her new habit of socializing:

> My husband used to go out a whole lot after work while I was home for seventeen years, and he would say, "Oh, we were talking

> business." I would just reheat his supper at 10 o'clock at night. Now since I've started working I've gotten in a crowd that invites me to go out with them after work. So, I started telling him, "I'm going to go out with the girls after work." Well, now it winds up that I go out more than he does. He's always telling me, "You're going out again?" I do go out a lot and I used to feel guilty, but I don't anymore. I feel like, he did it for seventeen years, and even if I do it for more than seventeen years, I'm still doing what the generation does now.

For Donna, "going out with the girls" is a simple pleasure that she enjoys as a fringe benefit of being in the paid workforce. But it is also a way of renegotiating the traditional gender relations that characterized the first half of her marriage. She is claiming her right to an independent, public life that is separate from her husband and her family, and that is consistent with the greater autonomy for women that is "what the generation does now."[16]

Personal transformations do not occur without resistance, however, from both husbands and from the community at large. Perhaps the most painful realization for wives in the Sparrows Point area who have challenged traditional gender relations is that they will have to disappoint someone. After generations of accommodating husbands' work schedules and needs for domestic services, the right of women to engage in independent activities outside of their homes and away from their families has been one of the most difficult mindsets that women have had to confront, both in their husbands and in themselves. Pauline Pearson found that simply asserting the right to do things outside her home for her own enjoyment was a big step: "The day I left the dirty breakfast dishes in the sink and went jogging with a friend was the beginning of a whole new way of thinking for me. What an idea, realizing that it wasn't bad that I wouldn't be there if somebody needed a ride to the store or if my husband came home looking for something to eat."[17]

Pauline took the next step when a group of women friends decided to spend a week at the beach without their families. Once she overcame her own reluctance to go she had to face the protests of her husband:

> The first time I went to Ocean City with my girlfriends . . . my husband was hurt and angry. He couldn't understand why I would even *want* to go anywhere without him. The day we left I put my bags by the front door, assuming he would carry them to the car

> for me, as he always does. He said, "If you can go to Ocean City by yourself, you can carry your bags by yourself." He didn't talk to me for a week after I got back.

For Pauline, the important thing was that she faced her husband's disapproval and went to Ocean City with her friends despite his objections. She saw the event as the beginning of her own independence, but also as a turning point in her marriage. Henceforth she could challenge the prevailing assumptions about a wife's duty to her husband. Her husband might protest, but the marriage would survive the change.

Censure of women who travel or vacation without their families comes not just from husbands, but also from the prevailing community ethic. Devotion to family is a closely held value in the neighborhoods near the Point, and as women redefine their family arrangements they face criticism from other women. When Gail Dobson decided to go on a spiritual retreat in Chicago with a female friend, her husband urged her repeatedly to stay home, but it was other women at her workplace who registered moral objections to what they perceived as the abandonment of her family responsibilities: "Even the day before I was leaving, my husband said, 'Why don't you forget about Chicago and stay home?' But it was my female coworkers who gave me the most grief. They kept saying, 'How can you leave your family like that?'" As Gail proceeded to redefine the gender roles within her marriage, she discovered that her campaign for more autonomy had to be waged not only with her husband, but also with other women who aligned themselves with the more traditional forces in the community.[18]

The wives I interviewed who have gone into the paid workforce have developed female friendships outside of their family circles and away from their neighborhoods, friendships that are based on common interests and work-related ambitions rather than on traditional family responsibilities. The experience of having friendships that are work-related and independent of their families is a phenomenon that makes an expanded social world accessible. Women in this community have formed support networks that are based in the workplace and that encourage more personal independence and more egalitarian marriages, and they are exposed to a public consciousness that broadcasts more self-determination for women. Wives in this steelmaking community can recite the reprimands of relatives, friends, or neighbors who have chastised them for breaking with the tradition of the homebound steelworker wife, and contemporary women who are in the paid workforce rely on support networks of like-minded female relatives, neighbors, friends, and coworkers who are also wrestling with traditional gender roles.[19]

Negotiating the shared responsibility for childrearing is the single most noticeable change to have occurred in the communities surrounding the Sparrows Point steel mill. Deindustrialization has released men from the steel industry's swing shifts and unexpected double shifts, pushed women into the full-time paid workforce, and compelled many couples to arrange their work schedules so that child care is shared. There are couples that rely on a relative or a day care center to care for young children, but not all families are willing or able to choose those options. Jacqueline Wallace described an arrangement between her husband and herself that met both their work schedules and took into account their various skills as well as likes and dislikes: "When I went to work we were able to arrange it so our children had a minimum of time with a babysitter. Pete's schedule was such that he could take care of them a lot, and he did everything—get them ready, get them breakfast, get them lunch, even dinner sometimes. It wasn't below him to do laundry or iron. If the two of us were going to work we had to do it that way."

Jacqueline's experience coincides with sociologist Arlie Hochschild's study, *The Second Shift* (1989), which argues that statistically men in blue-collar communities contribute somewhat more to shared household responsibilities than men with professional jobs. Part of the reason for that in the communities surrounding the Sparrows Point steel mill has to do with the economic history of the area. Household financial well-being has, since 1887, relied on the flexible strategies of both wives and husbands for sharing responsibilities. Just as wives took in boarders at the turn of the twentieth century when it was necessary to bring additional income into the household, at the turn of the twenty-first century husbands assume responsibility for child care as a practical solution to an economic problem. Because hiring child care is simply too expensive for many two-paycheck families, husband and wife work out creative—and often complex—plans for staggering working hours and child care.[20]

In neighborhoods adjacent to the Point, conversations among women about their husbands and their marriages have become conversations about partnership, how to build partnerships, and how to make them work, and women are proud of the ways in which their marriages work as partnerships. Women describe their marriages as being based not on abstract concepts such as equality as much as on joint commitments to shared goals that include financial contributions as well as household responsibilities.

Having a say in the purchase of a car is important to Pamela Baneck as the basis for a partnership: "Jim and I are partners. We are able to be partners because I have always handled the finances and neither one of us makes financial decisions alone. We always make them together. I'll never forget the weekend Jim came home with a jeep and I said I didn't like the canvas doors. On Monday he took the jeep back and got another car." In this case Pamela saw her veto power over the kind of car her husband bought as a sign of their partnership, despite the fact that she didn't go with him to the car dealership to participate in the actual shopping. It was sufficient for Jackie that her objections were taken seriously by her husband: "We both work, but I keep the checkbook. Once a week we sit down and say, 'This is what we want.' Then we figure out how we work these things into our budget. We always talk about it first, and we never buy something unless we both want it."[21]

Pamela would be surprised at the number of women who similarly describe their husbands as partners, or who view their marriages as in the process of becoming partnerships. Theresa Parker had complaints about trying to share tasks at home with her husband, but she remained optimistic: "Women . . . need to take the initiative to show men that we are capable of doing the same job as they do, and do it just as well. Whether these jobs are in the workforce or in the household, we control more than we think we do. . . . We control our own destiny." In spite of resistance on the part of many former steelworkers, wives have been adamant about their desires to have marriages that are partnerships in some or all of the essential aspects of the relationship: handling the finances jointly, extending emotional support mutually, sharing childrearing cooperatively, and maintaining household business affairs as a team.[22]

Some couples have negotiated new sets of gender relations that are admirably egalitarian and affectionate. Pamela Baneck moved gradually over the past two decades from her role as full-time homemaker through a series of steps directed toward a successful administrative career. When asked in the late 1980s if there were things she continued to do to reassure her husband that he was not being left behind, she got a twinkle in her eye: "Oh yes. When we were first married I packed his lunch every day, and once in a while I would slip a love note into it. Now that I'm so busy, my husband packs his own lunch, but once in a while I still write him a love note, only now I sneak it into his shirt pocket." In interesting ways, this marriage has maintained a lot of its original structure of steelworkers in the 1950s and 1960s while improvising on the participation of both husband and wife.

The lunch still gets packed and the love notes still get sent, but the expectation that the wife will fix the lunch has vanished in light of her new responsibilities in the paid workforce. When I did a second interview with Pamela ten years later, her husband was getting ready to retire from his job, but she was continuing the administrative career that, despite increasing responsibilities and longer hours, she had grown to enjoy. "We would get in each other's hair if we were both home. So it's good that I'm still working. Besides, he does all the housework now. I've got it made." The Baneck marriage is representative of many couples that have successfully renegotiated their marriages from the breadwinner/homemaker gender roles of the 1950s and 1960s to a partnership between two breadwinners.

While some marriages evolve into strong partnerships, there are also marriages that become unhinged when a wife goes into the full-time paid workforce. Carolyn Bardzik talked about her perception that it was when she began earning more than her husband that their marriage began to erode: "When I got my job at SuperFresh I was making more money than George was. I was not in control of my destiny, but I had a great deal to say about what happened to me. It was then that my husband started to drink a lot more and we sort of drifted apart." When asked about community mores, Willa Martin, a woman whose family had been in Sparrows Point and Dundalk since 1900, recalled the prevailing attitudes in the 1950s and 1960s about divorce: "Of all those people that I knew, all of my relatives, those ten on one side and nine on the other, I never knew of anyone who was divorced. None of them. You just didn't."[23]

Willa's explanation for the absence of divorce in this steelmaking community prior to the 1970s is not that families were happier or that marriages were better than today. On the contrary, she speculated that the marriages in her family were deeply flawed:

> By today's standards my husband and I would think that both of our sets of parents were really dysfunctional if we compared them to our own standards for marriage, which is very different. Going back to my grandmother's marriage, there was no warmth there. There was no affection. That was just not a part of their ways. But they stayed together until they died. They were hard-working kinds of people, and the women were drudges in a lot of ways.

Despite the relative economic success of pre–World War II marriages, women who are in the paid workforce and seeking partnerships in their marriages refuse to accept the role of "drudge."[24]

If most steelworkers' wives prior to the 1970s accepted male behavior that by today's standards is objectionable, there was an occasional woman, like Jean Stackpole's mother, who put her foot down:

> When my mother first married, her husband was in the Aleutians during the war and she had an apartment and a job and she was self-sufficient. When he got home he wanted her to be dependent on him. He kept the paycheck and paid the bills and gave her an allowance for food and household expenses. If she ran out of milk and bread, he said, "Too bad, you have to be a better manager." This was intolerable to her.

When Jean's mother ended her marriage she faced the disapproval of her neighbors, because divorce lacked social sanction in the community: "I was seven in 1955 when my parents divorced and we went to live with my grandparents in Old Dundalk. My mother was a tainted woman, a Jezebel, for having divorced. There were people on our block who wouldn't let their grandchildren play with us because my mother was divorced." The taboo against divorce in the 1950s conformed to nationwide norms.[25]

If divorced women were considered Jezebels in the 1950s, there is now a different consciousness about women ending their marriages. Both men and women in these steelmaking communities are critical of men who drink, hit their wives, or run around with other women, and these male behaviors are considered sufficient justification for divorce. Indeed, women who put up with excessive drinking or bullying are often the objects of criticism for indulging male behavior that was excused a generation ago but is now considered unacceptable.[26]

The contemporary women I interviewed emphasized aspects of their divorces that represent their taking charge of their own lives. Earlier generations of Sparrows Point, Dundalk, and Turner Station residents emphasized job security and well-paid breadwinners as the source of the satisfaction in steelworker marriages. Edna James puts the emphasis elsewhere: "When I went through a divorce I learned to do things on my own. I learned how to make decisions, to support myself, and to raise a child. I learned how to do all the pieces." Although she is divorced, or maybe because she is divorced, her focus is not on a good income or a new boat, but on her accomplishments as a single mother: "Patrick got his Eagle badge in Scouts last year. Nothing could equal the pride I felt the night of the ceremony. It was the best thing I had ever done as a mother. When I remember all the parking lots I sat on, and all the miles I drove getting him to Scouts meetings. It was hard work. My mother was proud of me, too. My mother told the Scoutmaster, 'She's done this all by herself.'" Doing "all of the pieces" of her life is an apt metaphor for Edna's perception of her own success, on the basis of her achievements in the workplace as well as her sense that a divorce forced her to be self-reliant.[27]

The ways in which women in the neighborhoods near the Sparrows Point steel mill have challenged the patriarchal systems in which they grew up are varied. Some women have worked persistently to reorganize the division of labor in their households or to arrange the household finances in a way that reflects a more egalitarian husband-wife relationship. Other women have broken the community's taboo against divorce, left their marriages, and established independent, single lives.

These challenges to family systems of a generation ago that had a single breadwinner who was the "head of household" are all built upon small, everyday decisions and actions that take place quietly within individual homes, unnoticed by newspaper reporters or television cameras. Nonetheless, the efforts of wives to achieve meaningful changes in gender relations have transformed the dynamics within families in the residential areas surrounding the Sparrows Point steel mill.

Collectively these efforts have served to change the character of the communities near the Point in major ways. The man who takes his children to his mother's house while his wife runs the Bingo at church, the woman who vacations in Ocean City with her friends, and the couple that pays a $75.00 fee to put both names on the deed to the family home are all participants in a new set of gender relations that have been negotiated slowly, haltingly, and sometimes painfully between husbands and wives who may be unaware of or uninterested in the fact that their changing household arrangements are part of a larger sociological trend. Husbands often resist giving up traditional domestic entitlements at a time when their employment status and security are also under assault. But their resistance is more than matched, in most instances, by the tenacity and creativity with which women have carved out for themselves lives that push against the traditional constraints on female autonomy. Theresa Parker was realistic about the limits to her own ability to forge a partnership with her husband within her household, but she had high hopes for the next generation of women: "Daughters are being taught that they are capable of doing whatever they set out to do. Giving these children this confidence will provide them opportunities to have powerful positions in the workforce."

The renegotiation of gender roles involves issues that are directly linked to the transition from a single paycheck household to a dual-income household. The absence of an abundance of jobs at the Point is a major factor in the realignment of the structural and ideological underpinnings of family systems in neighborhoods nearby. Without the assurance of high-paying steelmaking jobs, and without the erratic demands of swing shift schedules, both men and women have flexibility in their roles that are conducive to changes in family systems. Wives agree that being in the paid workforce in permanent full-time jobs, while making the management of family life more complicated, has given them a degree of social power and personal independence unavailable to their mothers' generation. Wives also insist that changes in their marriages were achieved through negotiations that they have strategically pursued with their husbands. But men and women agree that it is the process of deindustrialization and the necessity of women once again contributing essential income to their families, which has spurred the redefinition of gender roles in the households with breadwinners and homemakers that characterized their community during the 1950s and 1960s.

8

A Larger Circle of Neighbors: Deindustrialization and the Web of Class, Race, Gender, and Location

My parents and their parents had only their own family and circle of neighbors. They felt the same way about things, they shared the same religious beliefs, and they were of the same economic group. They were just all alike. They lived in a really, really small world. Nowadays, we go out and we're in a diverse group and we have the opportunity to pick and choose what we see that appeals to us in people that have different thoughts about things, different beliefs, and different backgrounds.

—WILLA MARTIN, APRIL 1993

When the company town of Sparrows Point was constructed beginning in 1887, its design reflected a social hierarchy based on class, ethnicity, and race. The hierarchy was clearly marked with streets lettered A through K. There was little question of who belonged at the top of the hierarchy, who belonged at the bottom, or what place you and your family belonged anywhere in between.

The street plan established in the town of Sparrows Point in the late 1880s set in place for families of steelworkers a prevailing sense of having a circumscribed niche in the social world of the town. Sarah Wilson remembered that her mother, who had grown up in Sparrows Point, continued to embrace that sense of social limitations throughout her entire life:

> There were clear distinctions of who you were. Who you traveled with and where you lived said that to the whole world. You live on B Street and you were a something. And if you're not something, you don't live there. My mother's whole life she always used this phrase that I just hated, and I still hate to this day: "That is not for people like us," she would say. "That is *not* for people like us." My mother and her neighbors saw themselves as the sort of people who were not entitled to certain privileges.

The company town of Sparrows Point made clear demarcations between the professional class of doctors, teachers, and mill superintendents who lived on B Street and C Street, and the skilled white workers who lived north of D Street. Even the nuances of social class within the group of white steelworkers were marked by the distance between a neighborhood and the General Manager's house on B Street. The further north you lived, the lower your rank and pay was inside the mill. A different kind of demarcation in the Sparrows Point design was Humphrey's Creek, which placed the families of African American laborers across a physical boundary and at the bottom of the hierarchy. Because the black community of the company town was north of Humphrey's Creek, both blacks and whites referred to it as "the other side."[1]

Despite the race-based systems of residential segregation and job discrimination, between 1887 and World War II there existed a striking complement between black women and white women in Sparrows Point that was based on the widespread practice of taking in boarders. In both the white and the black sections of the company town, wives of steelworkers were providing domestic services for single male steelworkers who paid for room and board. Both groups were fixing meals, packing lunches, and laundering bedclothes to earn essential income for their households. Contemporary women can often remember whether or not their own mothers or grandmothers took in roomers or boarders.[2]

Unfortunately, the shared history of *both* black and white women cooking and cleaning for boarders or working in boardinghouses has not been acknowledged or made part of the community's consciousness. In the remembered history of race relations in the company town of Sparrows Point, the common experiences of both white and black women doing domestic work for boarders has been overshadowed by the practice of job discrimination in the steel mill, residential segregation in the company town, and Jim Crow practices in the town's stores and recreation facilities.[3]

FIG. 18 Turner Station and Dundalk women who are the wives, sisters, daughters, granddaughters, even great-granddaughters of steelworkers have crossed the barriers of segregation to work as activists on issues and organizations that affect both communities. Three women in this photo have family roots that go back at least as far as the 1900 census for the planned industrial community of Sparrows Point.

The parallel experiences of those white and black women who took in boarders has also been overshadowed by the practice of black women performing domestic service in the homes of white women who were married to the superintendents, foremen, and the highest-paid skilled workers. In *Domesticity and Dirt* (1989), Phyllis Palmer argues that the social identity associated with "whiteness" and "blackness" between 1920 and 1945 was based in part on African American women doing domestic work in the homes of white women. In this respect, race relations in Sparrows Point, and later in Turner Station and Dundalk, corresponded with American social practices of the time. Prior to 1974, white men held a monopoly on the higher-paying skilled jobs, while black men were limited to lower-paying laboring jobs at the Point. African American wives who needed to supplement the household income found work available as domestics in the homes of those white women who benefited financially from being married to steelworkers in skilled, higher-paying jobs.[4]

White women who grew up in the company town of Sparrows Point often have clear recollections of the African American women who came to their homes to do ironing and sometimes cleaning. Some particularly insightful white women with roots in the town of Sparrows Point are astute enough to understand that cleaning and ironing for white women was socially demeaning to black women. Mary Lipsett, a white woman whose siblings, mother, and grandmother all grew up in the town of Sparrows Point remembered that, although the subject of race was not openly discussed, race relations were a part of the social consciousness of white residents. Mary remembered that in her family's experience the town's class, race, and ethnic social arrangements were assumed to be as natural and unobjectionable as was the town's geographic design. Looking back with a contemporary consciousness, however, Mary reflected that African Americans did not find their separate and unequal status in Sparrows Point as acceptable as she had once assumed: "There was definitely discrimination, right up in your face. The black people lived on I Street and J Street. They had their own town over there and they all lived there—happily as far as I was ever aware. But today African Americans have a real, real different perception of that subject than we have."

African American women who discuss Sparrows Point, Dundalk, and Turner Station between 1920 and 1945 remember with resentment the work that their female relatives did. Jacinta Cole recalled her grandmother and great-grandmother experiencing the indignities that went along with working as domestics, including negotiating the white community of Dundalk as second-class citizens, as well as deferring to Jim Crow laws and customs. Jacinta emphasized the extent to which her female relatives tried to protect their children from racial affronts:

> My grandmother and great-grandmother both did cleaning in the homes of white women. They didn't talk a lot about their experiences around the children but they would talk to each other, and you got a very clear sense that there was a whole different world outside of the neighborhood that probably wasn't too pretty or too wonderful to have to experience. They worked really hard at us growing up not having to do what they did or go through what they had to go through.

For contemporary African American women descended from families whose roots go back to the company town of Sparrows Point, continuing to live and work in the communities adjacent to the Sparrows Point steel mill leaves them vulnerable to reminders of the domestic work that their female relatives did in white homes. Margo Lukas, an African American woman in her late forties who is well educated and has a professional job, found that as long as her employment was in the white steelmaking community of Dundalk she would be associated with the earlier generations of women in her family who had done domestic work: "My mother was raised on Sparrows Point. There were thirteen of them. They lived on two streets, I Street and J Street and maybe a few houses on K Street. They called it 'the other side.' My aunts and my grandmother did day work. Even today, elderly white women that I encounter as part of my profession will say to me, 'You look just like her, and she used to do work for me.'" Because the women in Margo's family are all quite attractive, the elderly white women may think that they are paying Margo a compliment, and they may remember their relationships with Margo's aunts and grandmother as congenial. But for Margo, the social identity that she has worked hard to establish as a professional is diminished in her own perception when she continues to be linked to the racial hierarchy that existed in Sparrows Point. The elderly white women who employed Margo's relatives lack the sensitivity to changing race relations necessary to relate to Margo as a successful professional rather than reminding her that she is the daughter and granddaughter of their own domestic servants. Margo commented on the inability of older white women from Dundalk to look back and reflect that the women in the Lukas family may have preferred to be cleaning their own homes and watching their own children.[5]

As the company town of Sparrows Point was demolished between 1954 and 1974, the black residents who relocated to Turner Station and the white residents who relocated to Dundalk moved into expanded social worlds. During World War II, Turner Station and Dundalk included newcomers—mostly from West Virginia and North Carolina—who had migrated to the area to work in war industries, and who created a shift in class relations. The World War II migration enlarged the "circle of neighbors" by adding poorer newcomers from rural areas to the well-established, cohesive steelmaking communities with their own set of stable community mores. Rural whites from Appalachia threatened the mores of the area, which had come to include the home ownership, rose gardens, and well-kept lawns that characterized the established communities of second- and third-generation steelworker families in both the black community of Turner Station and the white community of Dundalk. While the white newcomers from West Virginia were eventually able to assimilate into the Dundalk community, Turner Station continued to be segregated.[6]

Even after the demolition of the company town of Sparrows Point, the converging forces of segregated housing, schools, and churches imposed the status of second-class citizenship on the black families living in Turner Station. The irony of this system of segregation and discrimination was the fact that many black families in Turner Station held middle-class values, aspirations, and achievements that paralleled the values and accomplishments of white Dundalk families.[7]

During the post–World War II era, some Turner Station families were forced to be *more* determined that their children get good educations than were some of their white Dundalk neighbors. Black families in Turner Station knew that education was the only avenue for their children to escape the racial segregation and discrimination that frustrated earlier generations of black steelworkers and their wives. For white Dundalk families, the skilled jobs in the steel mill that were reserved for white men continued to be readily available until the 1970s. This made higher education unnecessary for those white families who felt assured of the prosperity that a skilled job in the mill could provide. For black families in Turner Station, upward mobility did not become available in the mill until the 1970s, precisely the decade when American steelmaking jobs began their precipitous decline. Families in Turner Station who wanted a better life for their children turned to education and geographic relocation.[8]

For children who grew up in Turner Station in the 1950s and 1960s, segregation had two faces. Within their community they felt protected and nurtured, but they remember the larger Dundalk community as foreign and sometimes hostile territory. A Turner Station resident remembered her young life in a community where children were deliberately shielded from the hurt or stigma that segregation might inflict:

> I have fond memories of growing up in Turners. I don't think there is any other place in the world where I could have been more comfortable. We could run around and play in the grass and watch the squirrels and throw acorns and skip rocks across the water and catch minnows and play with the grasshoppers. Everything was right there for us. We didn't go to Dundalk. We didn't go there to buy our clothes; our parents bought our clothes. We only went there to buy shoes. They did take us to try on shoes. We were sheltered by our parents from the rejection in Dundalk. If we went to an all-black function it was in the city.

This remembered history of the racial hierarchy between Dundalk and Turner Station, as well as the role that black women played as domestics a generation ago, affects young black women today. It has encouraged some African American women from Turner Station to seek professional jobs in other parts of the Baltimore metropolitan area in order to distance themselves from the Sparrows Point steelmaking area, where a history of white superiority/black subordination might stigmatize them and might affect their careers.[9]

When I asked black women from Turner Station if entering the full-time paid workforce was a vehicle for liberation from patriarchal family systems, they responded by insisting that their struggle was a different one. Those women from Turner Station who had already been employed full-time before the 1970s argue that getting *into* the paid workforce was not a vehicle for more independence, because, as Yvette Johnson said, when it comes to being able to work, "black women have *always been free.*" Moving *up* in the paid workforce is the struggle that Yvette wanted to discuss.[10]

African American women from Turner Station do not complain about being denied access to the paid workforce prior to the 1970s. For black women a more common complaint is that in the past they were denied advancement in the workplace because of race discrimination, and it is access to *better* jobs to which they aspire, just as black men in this community have struggled to gain access to skilled jobs in the steel mill: "Black women didn't have the same jobs as white women. In the hospitals they were nurses' aides, not nurses, or if they were nurses with degrees they still did a lot of the duties that the nurses aides did." Yvette Johnson, like other African American women who grew up in Turner Station, insist that it was not lack of skills and credentials that kept them out of the better jobs: "What kept them out was that they didn't have the experience. We had our degrees, but we took them home to our kitchens or back to our churches. The majority of black women were not in the ranks. We took the jobs that were available to us because we wanted to work and needed to work." Although Baltimore's was one of the first school systems to comply with the 1954 *Brown v. Board of Education* school desegregation decision, Baltimore County was slower to comply, and throughout the state of Maryland segregation in public accommodations and discrimination in many skilled and professional job categories persisted by law as well as by custom into the 1960s.[11]

Beginning in the 1970s, deindustrialization vastly enlarged the previously circumscribed location of both work and residence for both white and black families who had once lived and earned their living in a self-contained steelmaking area. The "Live, Work, Shop Dundalk" bumper stickers began disappearing from the backs of automobiles commuting to work in other parts of the Baltimore metropolitan area. The expanded geographic and social worlds that came with deindustrialization have resulted in a shift in the hierarchical social arrangements of the circumscribed location.[12]

Deindustrialization forced a further widening of the "circle of neighbors" as women from Dundalk and Turner Station went into the full-time paid workforce. Many of them began commuting to jobs in workplaces in the Baltimore metropolitan area that are outside of the steelmaking communities where their parents' generation lived, worked, shopped, and spent their leisure time. The location of new workplaces has forced women out of the tightly circumscribed geographic area of southeast Baltimore County, where the Sparrows Point plant is located.[13]

The expansion of women's opportunities and horizons that came with deindustrialization has also resulted in a shift in the relations of race and class. In the process of adjusting to that shift in hierarchies, black women and white women from these steelmaking communities have been able to experience once again some of the commonalities that they have shared since their great-grandmothers were accommodating boarders and their great-grandfathers were working twelve-hour shifts in the steel mill.

For African American women, commuting outside of the Turner Station/Dundalk area has meant freeing themselves from the local social strata, which for a century was one of white dominance and black subordination. While black women did not escape prejudice, racism, and discrimination, they did leave behind the traditional assumptions in parts of Dundalk's white community that blacks belonged in subordinate employment roles. For this reason, Jacinta Cole, like other women from Turner Station, has sought jobs in other parts of Baltimore County or in Baltimore:

> I went to college and then got a master's degree, but I looked outside of the Dundalk area for jobs because I didn't want to have to deal with preconceived notions of who I was, being from Turners. I work for the Johns Hopkins Mediation Program and I've done a lot of teaching. I've taught for Sojourner Douglass College and Morgan State University. Those were places where I knew my master's degree would be respected.

Leaving Turner Station opened the doors to opportunities that are unconstrained by the old racial hierarchies of the company town of Sparrows

Point, and in that respect Jacinta has benefited enormously as an individual even though she has had to leave behind a comfortable environment of neighbors and family.[14]

The geographic movement to new job opportunities in response to deindustrialization has also had a significant effect on white Dundalk women. Unlike black women from Turner Station, whose jobs in other parts of the metropolitan area allow them to move away from the experience of racial subordination in the Sparrows Point steelmaking area, white Dundalk women are experiencing prejudice against blue-collar communities as they commute to jobs outside of Dundalk. When deindustrialization forced Theresa Parker into an encounter with the "outside world," the consequences were unsettling:

> People who live in Dundalk define themselves as citizens in a different way from the stereotypes of Dundalk held by outsiders from other communities. When these outsiders hear the word Dundalk, their first comment is usually negative. We are often associated with terms such as uneducated, blue collar, white trash, fat slobs with no culture. This always pisses me off. I would like to think that there is a lot more to Dundalk than what these others make it out to be.

Class prejudices encountered in the workplace is often in a form that has been institutionalized in the management and marketing strategies of businesses in the Baltimore area. Donna Prince talked angrily about her first experience with class bigotry: "As a young person I worked for the phone company, and they used to call me in to high-level management meetings when they were deciding on promotional material. They'd say, 'If the girl from Dundalk can understand it, then anyone can; and if the mother of the girl from Dundalk can understand it, then we really know that anyone can understand it.'" This example of business professionals denigrating the intelligence of Dundalk women reveals the crudity with which white-collar workers made remarks in front of the person they were insulting. It is also a good example of bigotry that is based on both class and gender, since the insult suggests that women from Dundalk find it particularly difficult to understand complicated promotional material.[15]

Sally Frank, a woman who grew up in Dundalk and became a white-collar professional, came to realize that a blanket assumption by outsiders that Dundalk residents are "not that bright" could seriously affect the kind of financial services offered in that community:

> The central office in San Francisco told me, "You're in a blue-collar area and your customers are not that bright. They're not sophisticated and won't understand your product." I insisted on making this opportunity available in Dundalk, and made sure that we had a solid plan for explaining the annuity. I was really pleased when our sales were way above the average for other branches in the state. I felt that I proved something about blue-collar people.

There is evidence that it negatively affects other kinds of resources, including the kinds of consumer items offered in local stores and the kinds of public services provided by the county.[16]

Dundalk women told me accounts of beginning new jobs in other sections of Baltimore County and having someone comment that, "you don't look like you live in Dundalk." That prejudices about Dundalk are rampant in the white-collar workplace came as a shocking surprise to Jamie Smith, who said, "For the first time in my life I felt like a minority."[17]

Most damaging of all is the impact of these prejudices on educational planning and implementation. Educators are not exempt from the class biases that are pervasive throughout Baltimore County. At a Baltimore County teachers' in-service program, one teacher laughed when she heard that I was doing research on Dundalk, and offered the following joke: "'What is a Dundalk briefcase?' Answer: 'A six-pack of Budweiser.'" The joke's point is an obvious one: Dundalk residents are content to drink beer and do jobs that are not worthy of our respect. The form of the joke is identical to outmoded ethnic jokes or racist humor, but the focus has changed to target an industrial community. We want to dismiss the joke because it shows such a profound lack of understanding of the level of skill required to work in a steel mill. However, the joke conveys a dangerous message that industrial work is demeaning, and it discourages young people from training for occupational niches such as machinists, where there is a workforce shortage and opportunities for good pay.[18]

The ultimate injustice of the prejudices aimed at residents of Dundalk is that so much of the prosperity enjoyed by wealthier communities in Maryland is built upon the skill, hard work, and sacrifices made by steelmaking families for over a century, but overlooked by people whose inflated sense of social status is combined with a feeble or nonexistent sense of local history.[19]

Marcella Knowles, a Dundalk woman who attended a private college with an upper-middle-class student body, was dismayed to find prejudices about Dundalk widespread among her fellow students and her teachers. Marcella was the first person in her family to attend college, and she found herself perched precariously between her blue-collar neighbors and family members and the world of professional privilege that she was entering. Marcella was offended by students from more affluent families who made thoughtless comments: "The girls of the middle-class that I went to school with made offhand digs because I was from the working-class, paying for my own education."[20]

Open expression of prejudice based on class—rather than race or ethnicity—is condoned among these upper-middle-class students, who, in other respects, considered themselves thoughtful and polite. For the student from Dundalk the insults were quite personal: "They talk about the lower class as 'grits,' the guys that wear blue jeans and T-shirts and do messy work. Those are my neighbors." Just like many upwardly mobile people from industrial communities, the student from Dundalk has learned that it will be difficult for her to move into the world of professional and white-collar work without betraying her family, friends, and neighbors.[21]

Most disturbing is that the upper-middle-class students at this private college firmly believe that blue-collar workers are no longer needed in today's economy: "A girl in my sociology class told me that manual labor shouldn't be paid as much as white-collar work because manual labor is on its way out. She actually thinks that soon *computers will do everything*. The truth is, without blue-collar workers there wouldn't be any white-collar workers." Marcella was indignant that her entire community had been rendered obsolete by a group of people who had never set foot in an industrial community and who had no respect for the skills involved in or the importance of being a steelworker.[22]

The disrespect of upper-middle-class college students for people who perform manual labor was further demonstrated for Marcella by the relationships between her fellow students and the custodial staff that cleans their school: "The janitorial workers are looked down on by most of the students. To me they are just someone doing a job. I'm the only student I've ever seen say hello to the janitors. To me they are people who do jobs that are not that different from what my father does as an auto mechanic." The dissonance that Marcella experiences between the working lives of her family and neighbors in Dundalk, and the class-consciousness of some of the college-educated professionals she has encountered is painful. Her fellow students have been taught, from early in life, that they are different from and superior to the janitors who clean their school. Marcella has learned from growing up in a Dundalk family that the janitors at her college are "people who do jobs that are not that different from what my father does." In order to flourish in the white-collar world to which she is headed, she will have to either reject her class background or straddle two worlds that do not communicate easily across class boundaries.[23]

Marcella looked to the faculty of her college for some key to understanding class prejudice, only to receive a sadly disappointing message from one of the professionals credentialed as her guide and teacher: "Among the faculty, half of them greet the janitors and half don't. One of my education teachers gave a lecture about the need to teach lower class students morals and values. For her big example she told us that we need to teach students not to say, 'I have to take a piss,' like they do in Dundalk. I went up to the teacher after class and said, 'I'm from Dundalk and we don't say "piss."'" For a student attending a college outside of the Dundalk area with the goal of entering the professional workforce, these are disturbing stereotypes being expressed by students and instructors at an institution that is a twenty-minute drive away from the blast furnaces of Sparrows Point.[24]

Stereotypes about Dundalk are almost always associated with the fact that Dundalk is where steelworkers live, and they reveal a pervasive misunderstanding about how much Bethlehem Steel's Sparrows Point plant has automated its steelmaking operation and how extensively that has changed the nature of work in the plant as well as the characteristics of the Dundalk community. Today most Sparrows Point steelworkers are highly skilled, some with advanced training in electronics. Dundalk has evolved into a community of people with diverse occupations because employment at the steel mill is no longer readily available. Many retired steelworkers have retrained to be teachers, counselors, physical therapists, or business managers. My interviews in Dundalk brought me into contact with one woman who began her career as a secretary, continued to pursue her education until she earned a master's degree, and now teaches at the same private college where students told Marcella Knowles that Dundalk is where the "grits" live.[25]

The origin of the class prejudice directed at Dundalk has a lot to do with the geography of social relations. During the first half of the twentieth century, and, indeed, up until the 1960s, Baltimore was firmly rooted in an industrial economy that spawned large numbers of blue-collar neighborhoods filled with families with immigrant or migrant roots who were proud of their hardworking heritage. Since the end of World War II, the children of Baltimore's working-class immigrants had been moving to suburban communities in Baltimore County, a suburban county that encircles but is separate from Baltimore and where expansive lawns and suburban developments symbolize the achievement of upward mobility.[26]

Dundalk at one time embodied that suburban ideal. But the population in Dundalk has grown older, the prosperity attached to being a steelworker has diminished, and suburban sprawl has stampeded across all of the counties surrounding the Baltimore metropolitan area. Today, Dundalk is a blue-collar community in the midst of a county filled with suburban retreats for white-collar professionals and their families. The devaluation of industrial labor is attributable in part to the status anxiety among those suburbanites who are themselves only one or two generations removed from manual laborers.[27]

Some of those suburbanites from white-collar areas of Baltimore County chose to announce their prejudices publicly to thirty-year-old Jamie Smith from Dundalk:

> I was embarrassed one night when I was talking to some people in a bar in Towson, which is where I work. They asked me where I lived, and when I said 'Dundalk' they couldn't stop laughing and saying, "you're from *Dundalk?*" like it was a big joke. My brother is working in a beach resort at Ocean City this summer, and he won't tell anybody that he's from Dundalk. He just lies and says he's from somewhere else.

When there is danger that encounters with other people will result in Dundalk being viewed as a joke, there are some people who will simply lie about their origins. Ironically, this is quite different from the response of African Americans who grew up in Turner Station, achieved middle-class status, moved to the suburbs, and are proud of their origins.

To avoid being embarrassed by class bigotry, some Dundalk residents choose to remain within the confines of their own community, protected from the spoken or implied criticism of outsiders. This isolationist response is the exact opposite of the responses I heard from black women living in Turner Station who commute to areas outside of the Sparrows Point steelmaking area to avoid racial bigotry.[28]

Jacinta Cole is a black woman who has clearly benefited from commuting to work. She is also well aware that the process of large numbers of black women seeking white collar and professional jobs away from the Point has had a negative impact on Turner Station:

> The women who excelled, the women who got that education, the women who earned a good living, mostly and eventually moved out, and they didn't give back, which adds to the depletion of the community. It's like a double-edged sword. You've got to move on, in order to grow, and most women have to move out of Turners to do that. At the same time, what used to be a community with a lot of talented people and families who were able to help each other out, is now losing too many educated people.

As more and more upwardly mobile women—and men—have left Turner Station, the out-migration of educated, talented African Americans has left the community with an older, poorer population. Statistics archived by the Enoch Pratt Library system project that the population for Turner Station will get smaller, older, and poorer over the next ten years.[29]

In fact, Turner Station and Dundalk are both experiencing an out-migration of more prosperous families. Families leave Dundalk, not for lack of larger, nicer homes, but because of the lure of spacious lawns and a more rural atmosphere in the northeast suburbs of Harford County. The migration of families out of Turner Station and Dundalk continues a pattern that has been in place since the late 1800s, with families moving further and further out from the center of Baltimore. Population movement to Baltimore County accelerated after World War II with the widespread availability of automobiles and the development of a network of highways. Today suburbs that are even further removed from the urban center are attracting migrants from Baltimore County who can afford to move and who are willing to commute longer distances.[30]

Upwardly mobile black and white families and individuals are now moving to Turner Station and Dundalk from Baltimore for the same reasons that motivated earlier generations of migrants: better schools, nicer homes or apartments, more space for lawns and gardens, and an escape from Baltimore's inner city crime and congestion. The arrival of African American newcomers in significant numbers has generated race and class conflicts, partly because they are newcomers and partly because this is Dundalk's first experience with African Americans who are not restricted to a separate community. Furthermore, the new arrivals are poorer people who do not come from steelworker families that were able to move into the middle class on the foundation of unionized industrial wages. Consequently, the extensive homeownership enjoyed by Turner Station and Dundalk in the past is now competing with the rapid growth of rental properties that accommodate people who are transients rather than lifelong members of the community. Like the rest of America, both Turner Station and Dundalk began to have problems with drugs and crime in the 1970s, but the growth of apartment complexes as well as the migration of poorer people from Baltimore has increased those problems.[31]

Turner Station is better able than Dundalk to see the ways in which the problems generated by newcomers are problems related to class differences. The longtime black residents of Turner are accustomed to neighbors who are hardworking, church going, and community-minded. They view the newcomers as people who do not share their middle-class values. For whites in Dundalk there is confusion about whether the newcomers represent problems of class or problems of race. Older people in Dundalk explain the problems that are increasing in their community as issues of race and the increasing numbers of blacks in Dundalk. They do not always have the conceptual vocabulary to translate the changes they see into issues of class and a local economy with few industrial jobs.[32]

From the perspective of class differences, large numbers of prosperous middle-class blacks have left the area adjacent to the Sparrows Point steel mill and taken their considerable resources of skills and education to more suburban areas. Poorer blacks are moving in with few resources and with many of the problems of the urban poor. Furthermore, due to deindustrialization the newcomers are moving into a community that has lost 17,000 steelmaking jobs in the last twenty-five years.[33]

Nonetheless, as some prosperous families move out and some poorer families move in, there are women and men from both Turner Station and Dundalk who have chosen to stay in the communities where they grew up and where their families have been associated with the Sparrows Point steel mill for generations. Men have always taken leadership positions in local politics, and now women have joined them as community activists who are assuming leadership for the revitalization of their neighborhoods. To Betty Kelly, a woman who grew up in Dundalk and is raising her daughters in Dundalk, the community has been transformed: "Just look at all of the organizations and stores in Turner Station and Dundalk that are being run by women, starting with the Chamber of Commerce and so many of the community organizations. Women have taken a very active role in improving our community and keeping it a good place to live and raise families." For these women the decision not to move their homes or their work outside of these steelmaking communities is based on a belief in the overriding benefits of strong community ties. Women are active in and often the leaders of revitalization efforts that work with young people, record the history of the area, and prepare for its future. In particular, the Dundalk Renaissance Corporation is responsible for the redevelopment of an area in southeast Baltimore County that includes both Dundalk and Turner Station, has a woman in the position of executive director, and includes both white and black representatives in community advisory capacities. Likewise, women have assumed leadership positions in Project Millennium; the Turner Station Heritage Society; the Dundalk-Patapsco Neck Historical Society and Museum; and the Greater Dundalk Alliance.[34]

A significant number of white Dundalk women have a clear awareness that Turner Station was subordinated by virtue of race to the political clout and access to resources of the larger and more prosperous Dundalk community. Furthermore, many of these women work actively to build better race relations, either through organizations or through personal relationships. Karen Fields describes herself as a white Dundalk woman who is willing to drive through Turner Station at night to drop off black coworkers, something that is rarely done. She also devoted a great deal of time to teach a young, black, female coworker how to drive, and then she helped her through the process of getting a driver's license—one of the tools she needed to enter the teaching profession.[35]

For older people there has been a coming together of black and white at the Fleming Center, a new seniors' center that was built in Turner Station and that attracts white seniors from Dundalk because of the quality of its programs and the friendliness of the staff and participants. For younger people in their teens and twenties, racial barriers are less impenetrable because they are in schools and workplaces that actively challenge hierarchical divisions based on race. Multicultural education and the racially inclusive dynamics of popular culture have created a worldview in which cross-racial associations and friendships are possible. Young people from Turner Station and Dundalk are able to discuss race relations with a matter-of-fact vocabulary that eludes their parents' generation.[36]

Because dating and marriage across racial lines has increased since the 1980s, many adults in Turner Station and Dundalk now have biracial grandchildren. As a consequence, blacks, while still a minority of the population in the Sparrows Point area, are not clearly stigmatized as much as they were as a strange group from "the other side." Individual blacks and whites know more about each other's cultures, and they are more likely to be colleagues, friends, or quite possibly immediate family.

In ways that have been complex and that have taken a different form for black women and white women, deindustrialization has transformed the ways in which women from Turner Station and Dundalk experience both race relations and class relations. As women from Turner Station and Dundalk have gone into the full-time paid workforce, they are assuming new roles as employees who interact with one another as coworkers on jobs where old racial hierarchies of black subordination are leveled or reconfigured on the basis of skill. Many of the women from Turner Station and Dundalk who now participate through their jobs in a larger "circle of neighbors," have, as Willa Martin said, taken "the opportunity to pick and choose what we see that appeals to us in people that have different thoughts about things, different beliefs, and different backgrounds."

Conclusion

Changes occur because you go out and work. I think that makes the biggest difference. Dundalk and Turners Station women still keep house, raise children, and do the same things that our mothers did. But I think the thing that matters most is that a woman has a working life, a public, social life, where women like my mother never worked outside her house a day in her life after she was married.

—SARAH WILSON, AUGUST 2001

In the 1880s, during the era of America's most vigorous industrial growth, Frederick and Rufus Wood designed a steel mill and the company town of Sparrows Point, which eventually drew thousands of men and women to the Patapsco Neck region of Baltimore County. By the time of the 1900 census, the steel mill had brought 434 wives, mostly young and mostly from rural towns in Maryland, Pennsylvania, and Virginia, to provide meals, to clean laundry, and to make a home life for husbands, sons, fathers, brothers, and boarders in the town of Sparrows Point.

Studies of the steel industry have generally looked at steelmaking communities as though the experiences of men were primary. *Wives of Steel* attempts to balance that assumption by acknowledging that women have always been present in steelmaking communities and by asking who these women were and what they were doing. Men and women made steelmaking communities function through their participation in gender roles and gender relations that have continued to change through each era of their communities' development.

Turner Station and Dundalk are communities that in many respects reflect the issues that confront American women and men at the beginning of the twenty-first century. A radical change in the economic structure of this community has disrupted a way of life that had been set in place with deliberate design at the end of the nineteenth century. Just as industrialization sent men into the steel mills and assigned women the job of providing domestic services to that steelmaking paid workforce, deindustrialization has transformed those systems in significant ways that have affected women quite differently than men.

Looking at the nineteenth-century origins of Sparrows Point, Dundalk, and Turner Station reminds us that these communities, their way of life, and their family structures, were set in place not to accommodate women, not to accommodate men, and certainly not to accommodate families, but to accommodate the profitable production of steel. The patriarchal family structure that was characteristic of these communities fit the demands made by the steel industry for large numbers of men who would work long hours in the mill while women provided essential domestic services for that male workforce.

This assumption—that the community of Sparrows Point, and later Turner Station and Dundalk, existed primarily to accommodate the making of steel—was so basic to the social arrangements of community institutions that it was never questioned. Because steel caused Sparrows Point, Dundalk, and Turner Station to come into existence, and because steel once made the area relatively prosperous, there never seemed to be any reason to question how steel had influenced the organization of family life and the relationships between women and men in this community.

Nonetheless, gender played a crucial part in defining the social and economic parameters in communities in Patapsco Neck. Implicit in every workplace are understandings about gender that are built into the organization of work and that impinge upon the social relations among workers. For steelmaking, gender is writ especially large in part because steel is an industrial niche that was reserved for men for the better part of a century. Steel mills developed a male work culture, and an ethic of manliness flourishes within its domain.

Shift work is the other aspect of steelmaking that irremediably affects gender relations and the organization of family life. In the Sparrows Point steel mill, the organization of work around long hours and rotating shifts introduces time as an ingredient in gender relations, because, out of necessity, "the family works the schedule." The organization of domestic work that was assigned to women had to be arranged to accommodate the continually changing work schedules of male steelworkers. The swing shift became a constraint on family life by restricting men from being partners in their marriages and by making women "prisoners of the swing shift."

Even the relatively high wages that steelworkers brought home during the post–World War II era affected gender relations in ways that went unrecognized, perhaps even unnoticed, because of the worker/union/consumer ethic that higher wages equated to a better life. Steelworkers' ability to make good wages in exchange for long hours encouraged their wives to remain out of the paid workforce, first by making it important that they be at home for erratic swings in the shifts and second by making it unnecessary for them to contribute money to their households.

If steelworkers and their wives were not protesting this configuration of their work and domestic lives, it is primarily because the prosperity that steelmaking brought to this community overshadowed everything else. In step with the rest of America, Sparrows Point, Dundalk, and Turner Station believed that the most important thing was having jobs that earned them enough money to buy homes, cars, and consumer luxuries. "Those were good jobs" is a phrase that you still hear a lot in steelmaking communities. That phrase does not refer to a work schedule that was conducive to a shared, cooperative family life or a marriage based on equity and shared responsibilities. For steelworker families, sacrifices were made to accommodate the conditions that existed in the steel industry, where money could be made but only at the expense of time devoted to family life.

With the transformation of the steel industry over the past three decades, that value system is being redefined out of necessity. Losing steel jobs has hurt this community in measurable ways that should not be overlooked or ignored. There are indications that poverty—poor health care, transient rental populations, drug-related crime—is an issue confronting Turner Station and Dundalk for the first time, and, as one resident said, "This is what happens when you don't have good jobs." From the point of view that affluence is the greatest good that this community can achieve, we see only decline and loss.

But, the transformation of steel has not affected all families in the same way, nor has it affected all members of the family in the same way. For young men leaving high school, the decline of jobs in steel has been a major incentive to attend college or prepare for skilled jobs in other occupational niches. Young men in Turner Station and Dundalk who are unwilling or unable to enroll in higher education will face lifelong problems with unemployment and underemployment. Unlike earlier generations of high school graduates, young women face similar choices: train for a job or face an uncertain future. Without unlimited steelmaking jobs for men, access to the paid workforce will be based on skills rather than on gender.

High-paying jobs that required specialized skills but not higher education and that were accessible almost solely to men have been lost. But when the transformation of the steel industry is studied from the point of view of the new set of gender relations that is emerging with the entry of married women into the paid workforce, we see a more complicated picture of deindustrialization.

What changes have occurred in this steelmaking community now that in many households "a woman has a working life"? It is not surprising that Turner Station and Dundalk women who have entered the paid workforce full-time express a sense of power within their families and their communities that from their point of view was lacking in their mothers' lives. Repeatedly, women testify that earning a salary and bringing home a paycheck gives them a feeling of power to voice opinions, claim time for themselves, participate in decisions, and build partnerships with their husbands based on equality and mutual respect. Being in the paid workforce gives them access to public lives where they can establish a sense of self that expands their social worlds beyond the domestic, familial sense of self to which their mothers were limited.

Being in the paid workforce has also enabled women to challenge the traditional patriarchal family systems that characterized the communities of Sparrows Point, Turners Station, and Dundalk throughout their histories as steelmaking communities. Women who are contributing substantial amounts to the household finances are also in a position to justify greater decision-making roles in those households. Decisions about major purchases, the division of household tasks, and responsibility for child care are no longer rigidly divided between women and men in accordance with tradition-bound arrangements. All of these decisions are now open to discussion and negotiation.

Women attest to the deliberateness with which they engage in negotiations for a division of responsibilities within their families that they can justify as equitable. For each family the arrangement may be quite different, but women insist that equity rather than tradition be the guiding principle in designing these arrangements. They are strategic in their pursuit of change that, in their eyes, challenges tradition and achieves equity.

In the process of achieving more egalitarian marriages, women face both resistance and cooperation from husbands. The husband who mused that losing his job at Sparrows Point "might have been the best thing that could have happened to me" because it made him more available to his family, is expressing the most impressive achievement of more egalitarian gender and family relationships. The woman who took driving lessons behind her husband's back, and the woman who nagged her husband until he put her name on the deed to the family home, are both examples of women who strategically negotiated the resistance mounted by their husbands in order to achieve more equality in their marriages.

Negotiating more equitable gender relations is, in the eyes of women in Turner Station and Dundalk, an accessible goal and one that they have made measurable strides to accomplish. They see themselves as having moved forward, beyond an outmoded set of gender roles that restricted women to domestic services on behalf of a household full of weary steelworkers.

There are other obstacles to independence and autonomy that Turner Station and Dundalk women must confront. A long history of racial segregation forced the Turner Station community to build its own institutions, including churches, businesses, schools, and recreation programs. These powerful assets have served both cultural and self-help functions in Turner Station. The separation based on race of Turner Station and Dundalk, however, hinders residents from launching a truly unified campaign against gender and class inequities, and especially against common community problems like unemployment, drugs, and crime.

Here, again, the consequence of women being in the paid workforce has had a powerful influence on the relationships between black and white women. Rufus Woods' fastidious plan for racial separation in housing and employment did, in fact, maintain a tenacious hold for much too long, but because women have so much more of a public presence today, black women are no longer people who "have their own little town" that is located "on the other side." Pushed into the paid workforce by economic necessities, black and white women from Turner Station and Dundalk are now coworkers, bosses, and colleagues. They are working together in offices, banks, hospitals, factories, and retail stores. Their relationships with one another are more individualized and less bound by stereotypes than was ever possible as long as they did not interact with one another outside of their homes and neighborhoods except as housewives and domestic workers. Race may continue to confuse and confound, but in the workplace black and white women have business issues to discuss and resolve.

The pervasive class prejudice that is directed at their community from outside is much more difficult to confront than restrictive gender consciousness or outmoded race relations. Class bias is insidious in its effect on the social isolation of southeast Baltimore County. It affects political decisions that are made about the funding of education, and it creates a sense of negativity about the desirability of Turner Station and Dundalk as places to protect from urban blight. Most sadly, as in the case of the college students who are belittled for coming from Dundalk, it hinders young people from recognizing the intelligence and capabilities they have for education and jobs.

How did Turner Station and Dundalk change from being the idyllic tree-lined oases for both city and rural families that they were as late as 1960 to becoming communities that have to struggle to maintain their heritage of a strong work ethic and a sense of shared values? Much of the source of this problem comes from the influence of the automobile and suburbanization on the development of Baltimore County. When Sparrows Point, Dundalk, and Turner Station were in their heyday, they were home to families connected to local industries. Today, with a network of superhighways ringing their perimeter, the communities adjacent to the Sparrows Point steel mill are rarely the neighborhood of choice for steelworkers, most of whom are commuting, some from as far away as Pennsylvania.

Attracted to ever more distant suburbs, industrial workers have joined their professional cohorts in the trend toward moving ever farther away from workplaces, old neighborhoods, row homes, and a modest yard. The American Dream is no longer to take a step up to a better-paying industrial job in Sparrows Point, Dundalk, or Turner Station, but to take a giant leap away from workplace locations in order to enjoy the quiet and spacious acres of distant suburbs. But neighborhoods are not like trendy clothing fads, and it would be bad public policy for Marylanders to turn our collective noses up at communities that have worked so well for so long. Both Turner Station and Dundalk have long traditions of community activism that may make it possible to preserve what has taken so many decades to build: communities that are inviting to families and individuals who want pleasant living in a close-knit, cohesive neighborhood environment.

Undoubtedly, a noteworthy transformation has taken place in these steelmaking communities in the course of just thirty years. Both women and men in these communities have faced the challenges of declining employment at the Point, employment that had created their community and sustained it for more than a century. It is tempting to conclude that women are the unsung heroes in this social transformation. In many ways they have been the pioneers in moving into the paid workforce, in getting education and public roles, and in claiming equality within their households. It is essential to remember, however, that men who were accustomed to a powerful role as the family's sole breadwinner have also been able, perhaps in some cases with considerable coaxing, to share that role within their marriages.

NOTES

INTRODUCTION

1. For studies that examine the individual, family, and community experience of deindustrialization, see David Bensman and Roberta Lynch, *Rusted Dreams: Hard Times in a Steel Community* (New York: McGraw-Hill, 1987); Staughton Lynd, *The Fight Against Shutdowns: Youngstown's Steel Mill Closings* (San Pedro, Calif.: Singlejack Books, 1982); Cathy N. Davidson and Bill Bamberger, *Closing: The Life and Death of an American Factory* (New York: W. W. Norton, 1998); Thomas Dublin, *When the Mines Closed: Stories of Struggles in Hard Times* (Ithaca: Cornell University Press, 1998); Michael Frisch and Milton Rogovin, *Portraits in Steel* (Ithaca: Cornell University Press, 1993); Judith Modell, *A Town Without Steel: Envisioning Homestead* (Pittsburgh: University of Pittsburgh Press, 1998); and Gregory Pappas, *The Magic City: Unemployment in a Working-Class Community* (Ithaca: Cornell University Press, 1989).

2. Margaret Byington, *Homestead: The Households of a Mill Town* (1910; reprint, Pittsburgh: University of Pittsburgh Press, 1974); William H. Chafe, Raymond Gavins, and Robert Korstad, eds., *Remembering Jim Crow: African Americans Tell About Life in the Segregated South* (New York: New Press, 2001); Herbert G. Gutman, "Work, Culture, and Society in Industrializing America, 1815–1919," *American Historical Review* 78 (June 1973): 561; David Montgomery, *The Fall of the House of Labor: The Workplace, the State and American Labor Activism, 1865–1925* (Cambridge: Cambridge University Press, 1987).

3. Elizabeth Clark-Lewis, *Living In, Living Out: African American Domestics in Washington, D.C., 1910–1940* (Washington, D.C.: Smithsonian Institution Press, 1994); Barbara Fields, *Slavery and Freedom on the Middle Ground: Maryland During the Nineteenth Century* (New Haven: Yale University Press, 1985); Darlene Clark Hine, *Hine Sight: Black Women and the Re-Construction of American History* (Bloomington: Indiana University Press, 1994).

4. Byington, *Homestead*, xx; Louis S. Diggs, *From the Meadows to the Point: The Histories of the African American Community of Turners Station and What Was the African American Community in Sparrows Point* (Baltimore: Uptown Press, 2003), 185–214; Dundalk-Patapsco Neck Historical Society, "Reflections: Sparrows Point, Maryland, 1887–1976," booklet (Baltimore: Cavanaugh Press, 1973), 12–18. Note that the Dundalk-Patapsco Neck Historical Society changed its name to the Dundalk-Patapsco Neck Historical Society and Museum; all historical references to the organization—for instance, as author of this 1973 publication—will carry the old name, and all current references—for instance, photo credits—will carry the new.

5. Thomas Bell, *Out of This Furnace* (1941; reprint, Pittsburgh: University of Pittsburgh Press, 1976); John Bodnar, *Steelton: Immigration and Industrialization, 1870–1940* (Pittsburgh: University of Pittsburgh Press, 1977); David Brody, *Steelworkers in America: The Nonunion Era* (New York: Harper and Row, 1960); John A. Fitch, *The Steel Workers* (1910; reprint, Pittsburgh: University of Pittsburgh Press, 1989).

6. Byington, *Homestead*, 35–36, 171–72; Ruth Schwartz Cowan, *More Work for Mother:*

The Ironies of Household Technology from the Open Hearth to the Microwave (New York: Basic Books, 1983), 245; Darlene Clark Hine and Kathleen Thompson, *A Shining Thread of Hope: The History of Black Women in America* (New York: Broadway Books, 1998), 244–58; Modell, *A Town Without Steel*, 35–37; Susan Strasser, *Never Done: A History of American Housework* (New York: Pantheon, 1982), 145–61.

7. Byington, *Homestead*, 138–44; Modell, *A Town Without Steel*, 84. Maryland Steel Company Papers of Frederick W. Wood, Accession 884, Hagley Museum and Library, Greenville, Delaware (hereafter FWW Papers).

8. Stephanie Coontz, *The Way We Never Were: American Families and the Nostalgia Trap* (New York: Basic Books, 1992), 99; Steven Mintz and Susan Kellogg, *Domestic Revolutions: A Social History of American Family Life* (New York: The Free Press, 1988), 178–79.

9. Judith Stacey, "Can There Be a Feminist Ethnography?" in *Women's Words: The Feminist Practice of Oral History*, ed. Sherna Gluck and Daphne Patai, 11–120 (New York: Routledge, 1991).

10. Byington, *Homestead*, 171–84; Dennis Dickerson, *Out of the Crucible: Black Steelworkers in Western Pennsylvania, 1875–1980* (Albany: State University of New York Press, 1986); Tamara K. Hareven and Randolph Langenbach, *Amoskeag: Life and Work in an American Factory City* (New York: Pantheon, 1978); Beth Anne Shelton, *Women, Men, and Time: Gender Differences in Paid Work, Housework, and Leisure* (New York: Greenwood Press, 1992), 1; Joe W. Trotter and Earl Lewis, eds., *African Americans in the Industrial Age: A Documentary History, 1915–1945* (Boston: Northeastern University Press, 1996).

11. Bell, *Out of This Furnace*, 44–89; Robert Bruno, *Steelworker Alley: How Class Works in Youngstown* (Ithaca: Cornell University Press, 1999), 49–52; Byington, *Homestead*, 27–38, 109–13; Fitch, *The Steel Workers*, 110–13; David Halle, *America's Working Man* (Chicago: University of Chicago Press, 1984), 58–64; Laurie Mercier, *Anaconda: Labor, Community, and Culture in Montana's Smelter City* (Urbana: University of Illinois Press, 2002), 162; Modell, *A Town Without Steel*, 162; Pappas, *The Magic City*, 86–87; Roy Rosenzweig, *Eight Hours for What We Will: Workers and Leisure in an Industrial City, 1870–1920* (New York: Cambridge University Press, 1983), 35–48; E. Anthony Rotundo, *American Manhood: Transformations in Masculinity from the Revolution to the Modern Era* (New York: Basic Books, 1993), 225–27.

12. Karen Olson and Linda Shopes, "Crossing Boundaries, Building Bridges: Doing Oral History Among Working-Class Women and Men," in *Women's Words: The Feminist Practice of Oral History* (New York: Routledge, 1991), 189–204.

13. Ava Baron's study, *Work Engendered: Toward a New History of American Labor* (Ithaca: Cornell University Press, 1991), and Ruth Milkman's study, *Gender at Work: The Dynamics of Job Segregation by Sex During World War II* (Urbana: University of Illinois Press, 1987), provide a framework for viewing the masculine work culture at the Sparrows Point steel mill. For other studies that discuss the gendered nature of the workplace, see Susan Porter Benson, *Counter Cultures: Saleswomen, Managers, and Customers in American Department Stores, 1890–1940* (Urbana: University of Illinois Press, 1986); Mary H. Blewett, *Men, Women, and Work: Class, Gender, and Protest in the New England Shoe Industry, 1780–1910* (Urbana: University of Illinois Press, 1988); Patricia Cooper, *Once a Cigar Maker: Men, Women, and Work Culture in American Cigar Factories, 1900–1919* (Urbana: University of Illinois Press, 1987); and "Issue on Work Cultures," special issue, *Feminist Studies* 11 (Fall 1985).

14. U.S. Census Bureau, Twenty-Second Census (2000).

15. Diggs, *From the Meadows to the Point*, 168–69; Dundalk-Patapsco Neck Historical Society, "Reflections," 6.

16. For the theoretical underpinnings of this in-depth study of an identifiable steel community culture, see Clifford Geertz, "Thick Description: Toward an Interpretive Theory of Culture," in *The Interpretation of Cultures: Selected Essays* (New York, Basic Books, 1973). Also see Michael Frisch, *A Shared Authority: Essays on the Craft and Meaning of Oral and Public History* (Albany: State University of New York Press, 1990); Paul A. Shackel, *Memory in Black and White: Race, Commemoration, and the Post-Bellum Landscape* (Walnut Creek, Calif.:

AltaMira, 2003); Linda Shopes, "Oral History and the Study of Communities: Problems, Paradoxes, and Possibilities," *Journal of American History* 89, no. 2 (September 2002): 588–98.

17. For a more detailed discussion of my methodology, see Olson and Shopes, "Crossing Boundaries, Building Bridges," 190–91.

18. Byington, *Homestead*, 3–32; Modell, *A Town Without Steel*, 90–110; Bruce Nelson, *Divided We Stand: American Workers and the Struggle for Black Equality* (Princeton: Princeton University Press, 2001), 145–84.

19. Within the field of oral history, there are several theoretical and methodological studies of the relationship between human memory and personally and collectively constructed and reconstructed versions of the past. See, for example, David Blight, *Race and Reunion: The Civil War in American Memory* (Cambridge: Harvard University Press, 2001); Michael Kammen, *Mystic Chords of Memory: The Transformation of Tradition in American Culture* (New York: Vintage Press, 1993); Alessandro Portelli, *The Death of Luigi Trastulli: Form and Meaning in Oral History* (Albany: State University of New York Press, 1991); Carol Reardon, "Pickett's Charge: History and Memory," *Catoctin History* 2, no. 1 (Spring 2003): 6–16; and Roy Rosenzweig and David Thelen, *The Presence of the Past: Popular Uses of History in American Life* (New York: Columbia University Press, 1998.

20. The relationship between interviewers and their subjects is examined in Michael Frisch, *A Shared Authority: Essays on the Craft and Meaning of Oral and Public History* (Albany: State University of New York Press, 1990); Judith Okely and Helen Callaway, *Anthropology and Autobiography* (New York: Routledge, 1992); Kammen, *Mystic Chords of Memory*; and Blight, *Race and Remembrance.*

21. Bruno, *Steelworker Alley*, 14–15; Mike Davis, *Prisoners of the American Dream: Politics and Economy in the History of the U.S. Working Class* (New York: Verso, 1986), 136–38.

22. For a discussion of intersectionality and the concept that race and gender are mutually dependent, interlocking cultural constructions and projections, see Leslie McCall, *Complex Inequality: Gender, Class and Race in the New Economy* (New York: Routledge, 2001), 29–59; Richard Oestreicher, "Separate Tribes: Working-Class and Women's History," *Reviews in American History*, 19 (1991): 228–31; Stephanie J. Shaw, *What a Woman Ought To Be and To Do: Black Professional Women During the Jim Crow Era* (Chicago: University of Chicago Press, 1996), 250 n. 18; Valerie Smith, *Not Just Race, Not Just Gender: Black Feminist Readings* (New York: Routledge, 1998), xiii, 89.

23. Diggs, *From the Meadows to the Point*, 49.

24. John Bodnar, "Power and Memory in Oral History: Workers and Managers at Studebaker," *Journal of American History* 75, no. 4 (March 1989): 1201–21.

25. Linda Zeidman, "Sparrows Point, Dundalk, Highlandtown, Old West Baltimore: Home of Gold Dust and the Union Card," in *The Baltimore Book: New Views of Local History*, ed. Liz Fee, Linda Shopes, and Linda Zeidman (Philadelphia: Temple University Press, 1991), 186–88.

26. This dual perspective of pride and resentment is also discussed in Byington, *Homestead*, 46; Modell, *A Town Without Steel*, 33; and Hine, *Hine Sight*, 219–22, 242–45.

27. Bodnar, *Steelton*; Mercier, *Anaconda.*

28. It was the multifaceted perspectives of those interviews that allowed me to understand that gender, race, class, and age were interrelated factors that determined how members of steelworker families had differently experienced, remembered, and represented the process of deindustrialization. See McCall, *Complex Inequality*, 204–10; Olson and Shopes, "Crossing Boundaries, Building Bridges," 189–204; Smith, *Not Just Race, Not Just Gender*, viii, 89.

CHAPTER 1

1. Enthusiasm about the Sparrows Point mill vitalizing the economy of Maryland is found in Charles Hirschfeld, *Baltimore, 1870–1900: Studies in Social History* (Baltimore: Johns

Hopkins University Press, 1941). See also Brody, *Steelworkers in America*; Fitch, *The Steel Workers*; Richard M. Bernard, "A Portrait of Baltimore in 1880: Economic and Occupational Patterns in an Early American City," *Maryland Historical Magazine* 69 (Winter 1974): 341–60; Jessica I. Elfenbein, *The Making of a Modern City: Philanthropy, Civic Culture, and the Baltimore YMCA* (Gainesville: University Press of Florida, 2001); and William A. Whiteford, director, *Port Baltimore* (Baltimore: Maryland Public Television, 1993). For a comprehensive business and labor history of the Sparrows Point steel mill, see Mark Reutter, *Sparrows Point: Making Steel; the Rise and Ruin of American Industrial Might* (New York: Summit Books, 1988). See also Zeidman, "Sparrows Point, Dundalk, Highlandtown, Old West Baltimore." Additional information about the founding of the Sparrows Point mill—including photos, maps, and records of the history of the area prior to the founding of the mill and town—came from the archives of the Dundalk-Patapsco Neck Historical Society and Museum.

2. Board of World's Fair Managers, Maryland, *Maryland: Its Resources, Industries and Institutions* (Baltimore: Sun Job Printing Office, 1893), 445. Preliminary surveys were at once made, and a plant for the manufacture of red bricks to be used in the construction of the works was established in July of the same year. The foundations of the blast-furnace plant were begun in August 1887, and the first furnace was completed and blown in October 1889. Three others have since been completed. The furnaces are each eighty-five feet high, 22 ft. in diameter at the bosh, and have a daily capacity of from 225 to 300 tons when working on the usual mixtures of foreign ores.

The Baltimore and Sparrows Point Railroad, connecting these works with the Pennsylvania and the Baltimore and Ohio Railroad, was constructed beginning in March 1888, and opened for passenger and freight traffic in January 1889. The construction of the Bessemer plant and rolling mill was commenced in May 1889. The first heat of Bessemer steel, and the first Bessemer steel ever made in Maryland, was made at 4:17 P.M., August 1, 1891, and the first rail was rolled six days later. The capacity of the Bessemer Department as it now stands is from 1,800 to 2,000 tons per day, and the rolling mills about 1,500 tons per day.

3. See Byington, *Homestead.*

4. See U.S. Census Bureau, Twelfth Census (1900); and Reutter, *Sparrows Point,* 62–65. For a discussion of the recruitment of the labor force in Pennsylvania steel mills, see Bodnar, *Steelton.*

5. U.S. Census Bureau, Twelfth Census (1900); also see the discussion of African American steelworkers in Dennis Dickerson, *Out of the Crucible: Black Steelworkers in Western Pennsylvania, 1875–1980* (Albany: State University of New York Press, 1986); John Hinshaw, "Dialectic of Division: Race and Power Among Western Pennsylvania Steelworkers" (Ph.D. diss., Carnegie Mellon University, 1995); Nelson, *Divided We Stand*; Joe William Trotter, *The Great Migration in Historical Perspective: New Dimensions of Race, Class, and Gender* (Bloomington: Indiana University Press, 1991).

6. U.S. Census Bureau, Twelfth Census (1900); Reutter, *Sparrows Point,* 64–65.

7. It was the Western Maryland county of Allegheny that had the highest percentage of foreign-born residents in Maryland in 1890, due to the recruitment of miners and laborers for the mines in that area. Sparrows Point's reliance on native-born white and black workers continued until the World War I era. Board of World's Fair Managers, *Maryland,* 445.

8. *The Union,* August 1892.

9. Records of the development of the steel mill and the town of Sparrows Point between 1887 and 1907 are found in the letters of Frederick and Rufus Wood, which are archived in the FWW Papers. See also U.S. Bureau of Labor, *Housing of the Working People in the United States by Employers,* Bulletin 54 (September 1904): 1214–15.

10. Dundalk-Patapsco Neck Historical Society, "Reflections."

11. Studies of nineteenth- and twentieth-century industrial towns and the systems used for creating an orderly labor force include Byington, *Homestead*; Stanley Buder, *Pullman: An Experiment in Industrial Order and Community Planning, 1880–1930* (New York: Oxford University Press, 1967); Thomas Dublin, *Women at Work: The Transformation of Work and*

Community in Lowell, Massachusetts, 1826–1860 (New York: Columbia University Press, 1979); Hareven and Langenbach, *Amoskeag.*

12. FWW Papers.

13. Marge Neal, "Sparrows Point: Well-Planned Homes Revealed Status by Proximity to Steel Plant," *Dundalk Eagle,* July 15, 1999.

14. For descriptions of Baltimore at the end of the nineteenth century, see William George Paul, "The Shadow of Equality: The Negro in Baltimore, 1864–1911" (Ph.D. diss., University of Wisconsin, 1972); Robert Brugger, *Maryland: A Middle Temperament, 1634–1980* (Baltimore: Johns Hopkins University Press, 1989); Fields, *Slavery and Freedom on the Middle Ground;* Mary Ellen Hayward and Charles Belfoure, *The Baltimore Rowhouse* (New York: Princeton Architectural Press, 2001); Cynthia Neverdon-Morton, "Black Housing Patterns in Baltimore City, 1885–1953," *Maryland Historian* 16 (Spring/Summer 1985): 25–39; and Karen Olson, "Old West Baltimore: Segregation, African-American Culture, and the Struggle for Equality," in *The Baltimore Book: New Views of Local History,* ed. Liz Fee, Linda Shopes, and Linda Zeidman (Philadelphia: Temple University Press, 1991).

15. Nelson, *Divided We Stand,* 160–64.

16. *Baltimore Sun,* October 21, 1906.

17. For a discussion of the lives of immigrants and African American steelworker families living in poverty at other mills during this era, see Bell, *Out of This Furnace,* 184–85; Byington, *Homestead,* 153–55; Cowan, *More Work for Mother,* 190; Fitch, *The Steel Workers,* 204; S. J. Kleinberg, *The Shadow of the Mills: Working-Class Families in Pittsburgh, 1870–1907* (Pittsburgh: University of Pittsburgh Press, 1989), xvii; and Nelson, *Divided We Stand,* 151–55; R. Douglas Hurt, ed., *African American Life in the Rural South, 1900–1950* (Columbia: University of Missouri Press, 2003).

18. Diggs, *From the Meadows to the Point;* Jerome R. Watson, *The Churches of Turners Station: A Legacy of Faith and Family* (Baltimore: Uptown Press, 2002), 67–69, 77–79; Jerome R. Watson, *Remembering Our Schools, A History of African-American Education in Turner Station and Sparrows Point* (Baltimore: Turners Station Heritage Foundation, 2004), 20–21; Dorothy Guy Bonvillain, "Cultural Pluralism and the Americanization of Immigrants: The Role of Public Schools and Ethnic Communities, Baltimore, 1890–1920" (Ph.D. diss., American University, 1999); and Leon Litwack, *Trouble in Mind: Black Southerners in the Age of Jim Crow* (New York: Random House, 1998): 142–78.

19. Charles Vert Willie, *Black and White Families: A Study in Complementarity* (Bayside, N.Y.: General Hall, 1985), 120–92.

20. Employment at the Sparrows Point steel mill waxed and waned according to economic conditions. However, the Bureau of Industrial Statistics of Maryland found that by 1901, the steel works at Sparrows Point, with four blast furnaces, a Bessemer plant and rail mill, and a complete steel shipbuilding plant, employed 2,000 men when it was running at full capacity. Bureau of Industrial Statistics of Maryland, *Fifth Annual Report of the Bureau of Industrial Statistics of Maryland, 1893–1901,* 21. By 1904, the U.S. Bureau of Labor reported that employment had increased to between 4,000 and 5,000 workmen. U.S. Bureau of Labor, *Housing of the Working People,* 1214–15.

21. The shanties more closely resembled the housing that Nelson describes for immigrants and African Americans during this era in New York, Pennsylvania, and Ohio. See Nelson, *Divided We Stand,* 154.

22. Watson, *Remembering Our Schools,* 19–21.

23. Watson, *The Churches of Turners Station,* 67–69.

24. Sandy Dunn, "Company," in *The Heat: Steelworker Lives and Legends* (Mena, Ark.: Cedar Hill Publications, 2002), 47–52; George L. Moore, "The Old 'Company Store' at Sparrows Point," *Baltimore Sun Magazine,* January 4, 1959. For a discussion of the impact of access to consumer goods on the lives of American women, see Lizabeth Cohen, *A Consumers' Republic: The Politics of Mass Consumption in Postwar America* (New York: Random House, 2003); Kathy Peiss, *Hope in a Jar: The Making of America's Beauty Culture* (New York: Henry

Holt, 1998). African Americans who lived in Sparrows Point describe the segregated company store in their neighborhood in Diggs, *From the Meadows to the Point*, 57.

25. Cowan, *More Work for Mother*, 234–38.

26. Dundalk-Patapsco Neck Historical Society, "Reflections."

27. FWW Papers; Robert Hessen, *Steel Titan: The Life of Charles M. Schwab* (Pittsburgh: University of Pittsburgh Press, 1990).

28. Zeidman, "Sparrows Point, Dundalk, Highlandtown, Old West Baltimore," 174–201.

29. Dundalk-Patapsco Neck Historical Society, "Reflections," 32.

30. Terri Narrell Mause, "A Community Built by Steel," *Dundalk Eagle*, March 1, 2000; Terri Narrell Mause, "Dundalk Timeline," *Dundalk Eagle*, n.d.

31. The Turner Station Heritage Society conducts ongoing research in the community of Turner Station under the leadership of Courtney Speed. Much of this research has been published in the work of historians Louis Diggs and Jerome Watson. The name of the African American residential community on the Patapsco Neck is sometimes spelled "Turner Station" and sometimes "Turners Station." Although "Turners" frequently appears in contemporary historical work, I use "Turner," which is the identification chosen by the community itself, the name currently used in the *Dundalk Eagle*, and the oldest recorded spelling. See also Paula R. Foltz, "Turner Station: A Community Moves Forward," unpublished manuscript, April 13, 2000.

32. For a discussion of the recreational activities of industrial workers at the turn of the twentieth century, see Paul Krause, *The Battle for Homestead, 1880–1892: Politics, Culture and Steel* (Pittsburgh: University of Pittsburgh Press, 1992), 164–65, 223–24; Rosenzweig, *Eight Hours for What We Will*, 85, 155–68, 173–76, 215. See also Robert Hilson Jr., "Osceola Smith Obituary," *Baltimore Sun*, November 23, 1997; Kweisi Mfume, *No Free Ride: From the Mean Streets to the Mainstream* (New York: Ballantine, 1996), 25–31.

33. Dundalk-Patapsco Neck Historical Society, "Reflections," 34. For a discussion of the impact of the Great Depression on steelworkers' wives, see Susan Ware, *Holding Their Own: American Women in the 1930s* (Boston: Twayne, 1982), 21–33; Lois Scharf, *To Work and to Wed: Female Employment, Feminism, and the Great Depression* (Westport, Conn.: Greenwood Press, 1980).

34. Reutter, *Sparrows Point*, 244–65. For discussion of the SWOC campaign for a union, see Nelson, *Divided We Stand*, 185–218; David Palmer, *Organizing Shipyards: Union Strategy in Three Northeast Ports, 1933–1945* (Ithaca: Cornell University Press, 1998); Vincent D. Sweeney, *The United Steelworkers of America* (Pittsburgh: United Steelworkers of America, 1956); Robert H. Zieger, *American Workers, American Unions, 1920–1985* (Baltimore: Johns Hopkins University Press, 1986); and Daniel Letwin, *The Challenge of Interracial Unionism: Alabama Coal Miners, 1878–1921* (Chapel Hill: University of North Carolina Press, 1998).

35. Kathi Ellington Dukes, "The Card," in *The Heat*, 36–38; Zeidman, "Sparrows Point, Dundalk, Highlandtown, Old West Baltimore," 186–87.

36. Zeidman, "Sparrows Point, Dundalk, Highlandtown, Old West Baltimore," 188; Dorothy Sue Cobble, ed., *Women and Unions: Forging a Partnership* (Ithaca: Cornell University Press, 1993).

37. Diggs, *From the Meadow to the Point*, 67; Nelson, *Divided We Stand*, 185–218.

38. Gail Porter Long, producer, *A Hometown at War* (Baltimore: Maryland Public Television, 1986); P. David Woodring, "Men of Steel, Hearts of Gold," in *The Heat*, 90–96.

39. "Advertisement for Day Village Homes, 1944," from the Dundalk-Patapsco Neck Museum of History and Culture archives.

40. U.S. Census Bureau, Twenty-Second Census (2000); Joe Nawrozki, "Remembering the Boom Times," *Baltimore Sun*, October 11, 1995, 1B, 4B.

CHAPTER 2

1. For studies of the lives of steelworkers, see Fitch, *The Steel Workers*; Brody, *Steelworkers in America*; Krause, *The Battle for Homestead, 1880–1892*; and Montgomery, *The Fall of the*

House of Labor. Several important studies of African American steelworkers have added to our understanding of working conditions in the steel industry, including Dickerson, *Out of the Crucible*; Hinshaw, "Dialectic of Division;" and Nelson, *Divided We Stand.*

2. Understanding how the organization of production in the steel industry affects family life in this steelmaking community requires an understanding of how steelworkers experience the day-to-day work regimen inside the Sparrows Point steel mill. This chapter is based on 28 interviews with men and women who were working at the Point or who had worked there. I began the interviews by asking people what it was like working at Sparrows Point. Invariably a description emerged of a physically intense, distinctly masculine work culture, from initiation rites and nicknames to the combativeness of interactions between coworkers and the constant presence of industrial hazards. An important insight into the cult of manliness at the Point was embedded in the different ways that women and men talked about the dangers in the mill. Women tended to discuss in detail a whole array of specific dangers that they encountered on the job and to be forthright about their fears. Men tended to mention dangers in the mill as a given, being careful never to appear afraid or to suggest that safety conditions should be improved. The guide I used most fruitfully to understand the work culture at the Point is Paul Willis's theoretical discussion of the interpretation of ethnographic studies of manliness and working-class culture in his article, "Shop Floor Culture, Masculinity and the Wage Form," in *Working-Class Culture: Studies in History and Theory*, ed. John Clarke, Charles Critcher, and Richard Honson, 185–98 (London: Hutchinson in association with the Centre for Contemporary Culture Studies, University of Birmingham, 1979).

3. Rufus Wood to Frederick Wood, April 21, 1890, FWW Papers.

4. Mercier, *Anaconda*, 177–78.

5. Rufus Wood to Frederick Wood, June 18, 1890, FWW Papers.

6. Rufus Wood to Frederick Wood, July 20, 1890, FWW Papers. Race, ethnicity, and kinship as a means of occupational entry into steel mills can be found in John Bodnar, *Workers' World: Kinship, Community, and Protest in an Industrial Society, 1900–1940* (Baltimore: Johns Hopkins University Press, 1982); and Joshua L. Rosenbloom, *Looking for Work: Searching for Workers: American Labor Markets During Industrialization* (New York: Cambridge University Press, 2002).

7. For a discussion of masculinity in working-class and industrial communities, see Gail Bederman, *Manliness and Civilization: A Cultural History of Gender and Race in the United States, 1880–1917* (Chicago: University of Chicago Press, 1995); Byington, *Homestead*, 109–13, 121; Nancy Hewitt, "'The Voice of Virile Labor': Labor Militancy, Community Solidarity, and Gender Identity Among Tampa's Latin Workers, 1880–1921," in *Work Engendered: Toward a New History of American Labor*, ed. Ava Baron (Ithaca: Cornell University Press, 1991), 142–67; Sean Wilentz, *Chants Democratic: New York City and the Rise of the American Working Class, 1788–1950* (New York: Oxford University Press, 1984); Willis, "Shop Floor Culture;" Michael Kimmel, *Manhood in America* (New York: Free Press, 1996); and Joseph Pleck, *The Myth of Masculinity* (Cambridge: MIT Press, 1983).

8. Working conditions at Sparrows Point at the turn of the twentieth century are described in U.S. Bureau of Labor, *Working Conditions and the Relations of Employers and Employees; from the Report on Conditions of Employment in the Iron and Steel Industry*, Document no. 110 (Washington, D.C.: Government Printing Office,1913).

9. Brody, *Steelworkers in America*, 170–71.

10. Brody, *Steelworkers in America*, 273–75; Rosenzweig, *Eight Hours for What We Will*, 39.

11. Brody, *Steelworkers in America*, 270–78; Bruno, *Steelworker Alley*; Richard M. Coleman, *Wide Awake at 3:00 A.M.: By Choice or By Chance?* (New York: W. H. Freeman and Company, 1986); Graham L. Staines and Joseph H. Pleck, *The Impact of Work Schedules on the Family* (Ann Arbor: Institute for Social Research, University of Michigan, Survey Research Center, 1983); and Halle, *America's Working Man*, 115–19.

12. Brody, *Steelworkers in America,* 122–26; Byington, *Homestead,* 37–40; and Halle, *America's Working Man,* 115–19.

13. Halle, *America's Working Man,* 117–22.

14. Halle, *America's Working Man,* 299. Byington, *Homestead,* establishes the long-range patterns of steel's affect on workers' families.

15. Brody, *Steelworkers in America,* 80–87; Bruno, *Steelworker Alley,* 77–78; and all of the following are from *The Heat*: Kathi Ellington Dukes, "Ma Beth," 69–71; Joe E. Gutierrez, "Missing at Work," 152–53; Gary Markley, "Who the Hell Are You?" 146–50; and J. A. Orellana, "Don't Mind the Noise," 128. The U.S. Bureau of Labor compiled statistics on accidents at Sparrows Point and at other U.S. steel mills and identified the most common causes of fatalities: burns, electric shock, falls, and being crushed to death.

16. Karen Beck Skold, "The Job He Left Behind: American Women in the Shipyards During World War II," in *Women, War, and Revolution,* ed. Carol Berkin and Clara Lovett (New York: Holmes and Meier Publishers, 1980), 60–65.

17. Kathi Ellington Dukes, "Remembering What's Important," 21–23, and Joe Gutierrez, "Snow Danced in August," 122–24, in *The Heat.*

18. Barbara Melosh, "Manly Work: Public Art and Masculinity in Depression America," in *Gender and American History Since 1890,* ed. Barbara Melosh (New York: Routledge, 1993), 155–89; Rotundo, *American Manhood,* 225.

19. Bruno, *Steelworker Alley,* 51; Gary Markley, "Darcy and the Silly Men," in *The Heat,* 73–89; Doreen Massey, ed., *Space, Place, and Gender* (Cambridge, U.K.: Blackwell, 1994), 193; Mercier, *Anaconda,* 122; Rotundo, *American Manhood,* 227.

20. Bederman, *Manliness and Civilization,* 20; Peter F. Murphy, *Studs, Tools, and the Family Jewels: Metaphors Men Live By* (Madison: University of Wisconsin Press, 2001). Several pieces in *The Heat* include accounts of the use of nicknames to tease and bully at the Point. See especially Stan Daniloski, "Wild Bill," 65–68; Markley, "Darcy and the Silly Men," 73–89; and Gary Markley, "Long Live 'D' Crew," 103–16.

21. Dukes, "Remembering What's Important," in *The Heat,* 21–23; Mercier, *Anaconda,* 162.

22. For a discussion of how competition and teamwork worked together, see Milton Keynes, "A Woman's Place," in Massey, *Space, Place, and Gender,* 193; Rotundo, *American Manhood,* 186.

23. For an overview of the history of alcohol use within working class groups of men, see Rosenzweig, *Eight Hours for What We Will,* 35–48.

24. Bederman, *Manliness and Civilization,* 17; Rosenzweig, *Eight Hours for What We Will,* 51–61; Markley, "Who the Hell Are You?" in *The Heat,* 145–46; Mercier, *Anaconda,* 28–29.

25. Byington, *Homestead,* 109–13, 121; Rosenzweig, *Eight Hours for What We Will,* 51–52.

26. Efforts to exclude black workers from skilled jobs in the steel industry are detailed in Hinshaw's *Steel and Steelworkers* and Nelson's *Divided We Stand.* Hinshaw argues that divisions between black and white workers in western Pennsylvania mills impeded class solidarity. Nelson focuses on steel mills in New York, Pennsylvania, and Ohio, and his comprehensive study highlights the ways in which Sparrows Point, as a steel mill in a border state, was unique in its employment, prior to World War I, of African American workers instead of southern and eastern European workers. Oral histories collected by Diggs in *From the Meadows to the Point* include several references to the attempts to exclude black workers from Sparrows Point; see especially 111–12. In the documentary *Struggles in Steel: A Story of African-American Steelworkers,* Tony Buba and Raymond Henderson interviewed black steelworkers from Sparrows Point and other mills about their efforts to gain access to skilled work at the Point. For analysis of discrimination against black workers in other types of industrial work, see Michael K. Honey, *Black Workers Remember: An Oral History of Segregation, Unionism, and the Freedom Struggle* (Berkeley and Los Angeles: University of California Press, 1999); Roger Horowitz and Rick Halpern, "Work, Race, and Identity: Self-Representation in the Narrative of Black Packinghouse Workers," *Oral History Review* 26, no. 1 (Winter/

Spring 1999): 23–43; William Harris, *The Harder We Run: Black Workers Since the Civil War* (New York: Oxford University Press, 1982); Howard Rabinowitz, *Race Relations in the Urban South, 1865–1890* (New York: Oxford University Press, 1978); Peter Rachleff, *Black Labor in the South: Richmond, Virginia, 1865–1890* (Philadelphia: Temple University Press, 1984); Charles Wesley, *Negro Labor in the United States, 1850–1925* (New York: Russell and Russell, 1927). In 1943, 7,000 of Bethlehem Steel's white workers walked out in protest over the company's plan to train fifteen black welders; see Sherry Olson, *Baltimore: The Building of an American City* (Baltimore: Johns Hopkins University Press, 1980), 364.

27. The concentration of African Americans in laboring jobs in the blast furnaces and coke ovens was common throughout the steel industry. In 1910, 73.6 percent of all black steelworkers in the United States were laborers compared to 8.2 percent who were skilled workers and 10.7 percent in the semi-skilled category. In 1935 in the Eastern district, which included Sparrows Point, 26.4 percent of blast furnace workers were black as compared to only 12.0 percent of the workers in the rolling mills. The statistical information on African-American men in the steel industry is from the U.S. Census Bureau, Thirteenth Census (1910); and Herbert R. Northrup, *Negro Employment in Basic Industry: A Study of Racial Policies in Six Industries* (Philadelphia: Wharton School/Industrial Research Unit, 1970), 258–59.

28. That union leadership was more liberal on issues of race than were the more conservative whites in the rank and file is explored in Zieger, *American Workers, American Unions*. Baltimore's black workers and their relationship with the CIO is documented in Roderick N. Ryon, "An Ambiguous Legacy: Baltimore Blacks and the CIO, 1936–1941," *Journal of Negro History* 65 (Winter 1980): 21–38; Davis, *Prisoners of the American Dream*; David R. Roediger, *The Wages of Whiteness: Race and the Making of the American Working Class* (New York: Verso, 1991).

29. Lawrence W. Levine, *Black Culture and Black Consciousness: Afro-American Folk Thought from Slavery to Freedom* (New York: Oxford University Press, 1977); Joe William Trotter, "African American Fraternal Associations in American History: An Introduction," *Social Science History* 28, no. 3 (Fall 2004): 355–66.

30. Herbert G. Gutman, *The Black Family in Slavery and Freedom, 1750–1925* (New York: Pantheon, 1976).

31. Vincent L. Matera, "Consent Decree on Seniority in the Steel Industry," *Monthly Labor Review* 53 (March 1975): 43–46; Jo Ann Ooiman Robinson, *Affirmative Action: A Documentary History* (Westport, Conn.: Greenwood Press, 2001), 185–88; Len Shindel, "They Acted Like Men and Were Treated Like Men," *Baltimore Sun*, February 17, 1992; Judith Stein, *Running Steel, Running America* (Chapel Hill: University of North Carolina Press, 1998).

32. Norman J. Brown Jr., "Dear Lisa," in *The Heat*, 132–33; Dukes, "Remembering What's Important," in *The Heat*, 22–23.

33. For a discussion of racial conflict in the workplace and its relationship to tensions at home, see Halle, *America's Working Man*, 299; and Mercier, *Anaconda*, 122. Several of the short stories in *The Heat* mention interracial friendships, including Jennifer Jones, "Day One," 27–30; Markley, "Darcy and the Silly Men," 73–89; and Markley, "Who the Hell Are You?" 149–50.

CHAPTER 3

1. See Jacqueline Jones, *American Work: Four Centuries of Black and White Labor* (New York: W. W. Norton, 1998); Alice Kessler-Harris, *Out to Work: A History of Wage-Earning Women in the United States* (New York: Oxford University Press, 1991); Lynn Y. Weiner, *From Working Girl to Working Mother, The Female Labor Force in the United States, 1820–1980* (Chapel Hill: University of North Carolina Press, 1985); Jo Ann E. Argersinger, *Making the Amalgamated: Gender, Ethnicity, and Class in the Baltimore Clothing Industry, 1899–1939* (Baltimore: Johns Hopkins University Press, 1999); Rosalyn Baxandall and Linda Gordon, *America's Working Women: A Documentary History, 1600 to the Present* (New York: W. W. Norton,

1995); Elizabeth Beardsley Butler, *Saleswomen in Mercantile Stores: Baltimore, 1909* (New York: Russell Sage Foundation, 1912); Hasia R. Diner, *Erin's Daughters in America: Irish Immigrant Women in the 19th Century* (Baltimore: Johns Hopkins University Press, 1983); Richard Hawkins, "The Baltimore Canning Industry and the Bahamian Pineapple Trade, c.1865–1926," *Maryland Historian* 26 (Fall/Winter 1995): 1–22; Roderick N. Ryon, "'Human Creatures' Lives': Baltimore Women and Work in Factories, 1880–1917," *Maryland Historical Magazine* 83 (Winter 1988): 346–64; Leslie Woodcock Tentler, *Wage-Earning Women: Industrial Work and Family Life in the United States, 1900–1930* (New York: Oxford University Press, 1979).

2. For a discussion of the lives of women and families in other steelmaking communities, see, for example, Kleinberg, *The Shadow of the Mills*; and Byington, *Homestead.*

3. U.S. Census Bureau, Twelfth Census (1900). Also see Byington, *Homestead,* 35–36, 171–72. The 1900 manuscript census differentiates between African American and white families. Information on the lives of African American women in Sparrows Point and in similar urban and industrial communities in the Upper South comes from Chafe, Gavins, and Korstad, eds., *Remembering Jim Crow*; Diggs, *From the Meadows to the Point*; Fields, *Slavery and Freedom on the Middle Ground*; Paula Giddings, *When and Where I Enter: The Impact of Black Women on Race and Sex in America* (New York: William Morrow, 1984); Gutman, *The Black Family in Slavery and Freedom, 1750–1925*; Hine, *Hine Sight*; Jacqueline Jones, *Labor of Love, Labor of Sorrow: Black Women, Work, and the Family from Slavery to the Present* (New York: Basic Books, 1985); and Trotter, *The Great Migration in Historical Perspective.* For a discussion of the differences between work done by black and white women during this period, see Jones, *American Work*; Kessler-Harris, *Out to Work,* 237–38; and David M. Katzman, *Seven Days a Week: Women and Domestic Service in Industrializing America* (New York: Oxford University Press, 1978).

4. Women whose husbands, sons, and boarders did shift work at the turn of the twentieth century reported that they often worked a twenty-four-hour shift because of men coming home from work at all hours. See, for example, Keynes, "A Woman's Place," 194.

5. U.S. Census Bureau, Twelfth Census (1900).

6. Taking in boarders was a common practice in working-class households during this period, but it was especially widespread in Sparrows Point because of the skewed ratio of single men to married men and because of the lack of adequate housing for single men in Sparrows Point at the turn of the twentieth century. Women who had boarders living in their homes were often serving food during most of the day, because men came home at irregular hours from working in the mill. See Byington, *Homestead,* 35–36, 171–72; Donna Gabaccia, *From the Other Side: Women, Gender, and Immigrant Life in the U.S., 1820–1990* (Bloomington: Indiana University Press, 1994); Hine, *Hine Sight,* 117–34; Modell, *A Town Without Steel,* 24–30; Valerie Kinkade Oppenheimer, *The Female Labor Force in the United States: Demographic and Economic Factors Governing Its Growth and Changing Composition* (Westport, Conn.: Greenwood Press, 1976), 233; and Strasser, *Never Done,* 159.

7. The 1930 census was the first in which the category of homemaker appeared. The secretary of commerce announced his intention "to give recognition to women in the home who hitherto, unless they had some moneymaking employment, were reported as having no occupation," as quoted in Annegret S. Ogden, *The Great American Housewife: From Helpmate to Wage Earner, 1776–1986* (Westport, Conn.: Greenwood Press, 1986), 159.

8. Studies of other working-class communities with boarders reveal both a smaller percentage of households with boarders as well as an older population of women taking in boarders. See Tamara K. Hareven, *Family Time and Industrial Time: The Relationship Between the Family and Work in a New England Industrial Community* (New York: Cambridge University Press, 1982); Myffannwy Morgan and Hilda H. Golden, "Immigrant Families in an Industrial City: A Study of Households in Holyoke, 1880," *Journal of Family History* 4 (Spring 1979): 59–68; Claudia Dale Goldin, "Family Strategies and the Family Economy in the Late Nineteenth Century: The Role of Secondary Workers," in *Philadelphia: Work, Space, Family,*

and Group Experience in the Nineteenth Century, Essays Toward an Interdisciplinary History of the City, ed. Theodore Hershberg, 277–310 (New York: Oxford University Press, 1981), 292.

9. Throughout the United States, 60 percent of all families lost at least one child in 1900. Peter Uhlenberg, "Cohort Variations in Family Life Cycle Experiences of U.S. Females," *Journal of Marriage and the Family* 36 (May 1974): 284–92.

10. Typescript statement of Maryland Steel Company, 1907, FWW Papers.

11. For a discussion of household labor at the end of the nineteenth century and before the availability of labor-saving appliances, see Glenna Matthews, *"Just a Housewife": The Rise and Fall of Domesticity in America* (New York: Oxford University Press, 1987); Mintz and Kellogg, *Domestic Revolutions*; Ogden, *The Great American Housewife*; and Shelton, *Women, Men and Time*.

12. To recover the everyday lives of steelworkers' wives between 1900 and 1945, I conducted in-depth ethnographic interviews with daughters and granddaughters who shared vivid memories of the generations who lived in Sparrows Point before World War II. The memories they recounted of mothers and grandmothers were filtered, of course, by the selectivity of memory itself, and in the accounts that people gave me they recalled the town of Sparrows Point through the lens of their own interpretations of the past. For this reason, I sought out people to interview from Sparrows Point, Turner Station, and Dundalk who were from diverse backgrounds in terms of age, sex, occupation, race, and ethnicity. I looked for patterns in the 80 interviews that I conducted, and I continued to conduct interviews over a fifteen-year period, between 1987 and 2002.

13. *Baltimore Sun*, October 21, 1906.

14. Quote from Mary Gorman in Zeidman, "Sparrows Point, Dundalk, Highlandtown, Old West Baltimore," 178.

15. The transformation of housewifery from a skilled craft to work perceived as easy is argued by Cowan, *More Work for Mother*, 245; and Matthews, *"Just A Housewife"*, 193. For a discussion of the changes taking place in women's lives during the 1920s, see Hine, *Hine Sight*, 270; and Sallie Westwood, *All Day, Every Day: Factory and Family in the Making of Women's Lives* (Urbana: University of Illinois Press, 1985).

16. The family size of Martin's grandparents was atypical of the 1920s, when birth rates dropped significantly. See David M. Kennedy, *Birth Control in America: The Career of Margaret Sanger* (New Haven: Yale University Press, 1970), 42–45. For a discussion of household work in industrial communities both before and after the installation of electricity and the availability of labor-saving appliances, see Cowan, *More Work for Mother*, 49.

17. For a discussion of the taboo against women in mines, steel mills, ships, and in other spaces determined to be masculine, see Byington, *Homestead*, 64; Brody, *Steelworkers in America*, 26; Dublin, *When the Mines Closed*, 38; and Mercier, *Anaconda*, 82.

18. On the work done by women and their children, see Tamara K. Hareven, *Family Time and Industrial Time: The Relationship Between the Family and Work in a New England Industrial Community* (New York: Cambridge University Press, 1982); Mintz and Kellogg, 27.

19. Byington, *Homestead*, 47.

20. Letter to Charles Kurtz from friends living in the company town of Sparrows Point, Maryland, March 24, 1934. Also see Dundalk-Patapsco Neck Historical Society, "Reflections," 23–24.

21. For a discussion of working-class wives and their roles as financial managers, see Byington, *Homestead*, 60, 108, 154.

22. John Modell and Tamara K. Hareven, "Urbanization and the Malleable Household: An Examination of Boarding and Lodging in American Families," *Journal of Marriage and the Family* 35 (August 1973): 467–79.

23. For a discussion of women's expanded economic roles in industrial communities during the Great Depression, see Ware, 8; and Cowan, 181–90.

24. The complexities of women taking in single male boarders are discussed in Byington, *Homestead*, 114; Gabaccia, 135–42. Strasser discusses the extent to which women who took in single male boarders became surrogate mothers for them. See page 155.

25. The trend toward increasing numbers of working mothers began nationwide in the 1920s, but it was generally met with disapproval. In Sparrows Point, married women continued to take in boarders, but widows had expanded their roles. See Frank L. Hopkins, "Should Wives Work?" *American Mercury* 39 (December 1936): 414–15; Leslie Woodcock Tentler, *Wage-Earning Women: Industrial Work and Family Life in the United States, 1900–1930* (New York: Oxford University Press, 1979), 138; Susan Ware, *Holding Their Own: American Women in the 1930s* (Boston: Twayne, 1982).

26. For a discussion of the persistence into the twentieth century of women and children working in industrial communities, see Tamara Hareven, "Review Essay: Origin of the Modern Family in the United States," *Journal of Social History* 17 (Winter 1983): 341–42; and Steven Ruggles, "The Origins of African-American Family Structure," *American Sociological Review* 59 (February 1994): 136–51.

27. By the 1930s, 50 percent of American households owned an automobile. Lizabeth Cohen, *A Consumers' Republic: The Politics of Mass Consumption in Postwar America* (New York: Random House, 2003), 173. Ethnographic information indicates that in Sparrows Point it was unheard-of for a married woman to even drive a car and rare for a widow to own one. In his biography, *No Free Ride: From the Mean Streets to the Mainstream* (New York: Ballantine, 1996), Kweisi Mfume remarks that in Sparrows Point in the 1940s and 1950s women did not drive.

28. On domestic service during this period, see Susan Strasser, *Never Done: A History of American Housework* (New York: Henry Holt, 1982); and David M. Katzman, *Seven Days a Week: Domestic Service in Industrializing America* (Urbana: University of Illinois Press, 1978).

29. Linda Zeidman, "Sparrows Point, Dundalk, Highlandtown, Old West Baltimore: Home of Gold Dust and the Union Card," in *The Baltimore Book: New Views of Local History*, ed. Liz Fee, Linda Shopes, and Linda Zeidman (Philadelphia: Temple University Press, 1991), 183.

30. The system of boarding houses and rooming houses, as well as the practice of wives taking in boarders, continued into the 1960s. The percentage of steelworkers boarding with a family or in a boarding house began declining, however, once government- and company-subsidized housing proliferated during World War I.

31. In 1870, 50 percent of employed women were domestic servants. This changed after 1900 because of the decline of immigration and the increase in pay for domestic work. Cowan, 120.

CHAPTER 4

1. The Sparrows Point steel mill reported 30,000 employees in 1950. For a discussion of the employment of married women prior to the 1960s, see Kessler-Harris, *Out to Work*; Oppenheimer, *The Female Labor Force in the United States*, 233; Tentler, *Wage-Earning Women*; and Winifred Wandersee, *Women's Work and Family Values, 1920–1940* (Cambridge: Harvard University Press, 1981).

2. For a discussion of American affluence in the 1950s and 1960s, see Steven Mintz and Susan Kellogg, *Domestic Revolutions: A Social History of American Family Life* (New York: Free Press, 1988); and David M. Potter, *People of Plenty: Economic Abundance and the American Character* (Chicago: University of Chicago Press, 1954).

3. See Matthews, *"Just a Housewife"*, 211. Matthews determines that the popular literature of the 1950s shares the view that a "'normal,' feminine woman would be happy staying at home. One who was unhappy was, in fact, by definition, not normal." See also Elizabeth Clark-Lewis, *Living In, Living Out: African American Domestics and the Great Migration* (New York: Kodansha International, 1996); Susan Hartmann, "Women's Employment and the Domestic Ideal in the Early Cold War Years," in *Not June Cleaver: Women and Gender in Postwar America, 1945–1960*, ed. Joanne Meyerowitz (Philadelphia: Temple University Press, 1994), 89–98; Joy Parr, *The Gender of Breadwinners: Women, Men, and Change in Two Indus-*

trial Towns, 1880–1950 (Toronto: University of Toronto Press, 1990), 104; Arlene S. Skolnick, *Embattled Paradise: The American Family in an Age of Uncertainty* (New York: Basic Books, 1991), 52; and Susan Strasser, *Never Done: A History of American Housework* (New York: Henry Holt, 1982).

4. See Claudia Goldin, *Understanding the Gender Gap: An Economic History of American Women* (New York: Oxford University Press, 1990), 97–98; and Barbara Ehrenreich and Deidre English, *For Her Own Good: 150 Years of the Experts' Advice to Women* (Garden City, N.Y.: Anchor Press, 1979). In 1930, according to Goldin, the rate of participation of married white women in the paid workforce was less than 10 percent. Married black women were much more likely to be in the paid workforce, but as a group black women were concentrated, even as late as 1970, in the lowest-paying jobs: "In 1890, 92% of all employed black women were agricultural workers and servants; in 1930, 90% were; and even by 1970, 44% were."

5. James Cunningham and Nadja Zalozar, "The Economic Progress of Black Women, 1940–1980: Occupational Distribution and Relative Wages," *Industrial and Labor Review* 45, no. 3 (1992): 540–55. On African American women professionals, see Darlene Clark Hine and Kathleen Thompson, *A Shining Thread of Hope: The History of Black Women in America* (New York: Broadway Books, 1998), 221–24, 263; Darlene Clark Hine, "'They Shall Mount Up with Wings as Eagles': Historical Images of Black Nurses, 1890–1950," in *Images of Nurses: Perspectives from History, Art, and Literature,* ed. Anne Hudson Jones (Philadelphia: University of Philadelphia Press, 1988), 189; and Natalie Sokoloff, *Black Women and White Women in the Professions* (New York: Routledge, 1992). Bart Landry, *Black Working Wives: Pioneers of the American Family Revolution* (Berkeley and Los Angeles: University of California Press, 2000), explores the consequences of a large number of black women having entered the paid workforce eight decades before a significant number of married white women worked outside the home.

6. Matera, "Consent Decree on Seniority in the Steel Industry."

7. The ways in which the lives of black women in industrial communities were parallel to the lives of white women is presented with both statistical analysis and ethnographies in Willie, *Black and White Families,* 120–87. For a discussion of the varying notions of "community" among different groups of women, see "Communities of Women," special issue, *Signs: Journal of Women in Culture and Society* 10 (Summer 1985); Nancy A. Hewitt, "Beyond the Search for Sisterhood: American Women's History in the 1980s," *Social History* 10 (October 1985): 306–7; Hine and Thompson, *Shining Thread of Hope,* 213–15; and Jacquelyn Jones, "One Big Happy Family?" *Women's Review of Books* 6 (February 1989): 4–6.

8. The photos taken in Turner Station included in this book provide a visual statement of well-dressed African American wives of steelworkers. These images reflect the unusually high standard of living for black families in Sparrows Point and Turner Station.

9. Historians of housework argue that after World War II American women, assisted by appliances, increased the time spent doing housework as a way of proving that they contributed to the household income.

10. See Strasser, *Never Done,* 268. Strasser argues that automatic washers and driers transformed laundry from a "weekly nightmare to an unending task" because the convenience of these appliances encouraged family members to throw their clothes in the hamper more often. The effect of a consumer culture on working Americans is analyzed in Lawrence B. Glickman, *A Living Wage: American Workers and the Making of Consumer Society* (Ithaca: Cornell University Press, 1997).

11. For a discussion of whether industrial communities were working class or middle class in the 1950s, see Diggs, *From the Meadows to the Point,* 100; Dundalk-Patapsco Neck Historical Society, "Reflections," 32–35; Rosenzweig, *Eight Hours for What We Will,* 173–76.

12. Bruno, *Steelworker Alley,* 84; Cohen, *A Consumers' Republic,* 232; Pappas, *The Magic City,* 29–30; and Potter, *People of Plenty,* 5–13.

13. The classic accounts of steelmaking communities all remark on the tension within families caused by shift work. See Bell, *Out of This Furnace*; Byington, *Homestead,* 32–36; and Fitch, *The Steel Workers,* 12–17.

14. Byington, *Homestead*, 37, 171–72; Halle, 107.

15. See Arlie Hochschild, *The Second Shift: Working Parents and the Revolution at Home* (New York: Viking, 1989). Hochschild argues that when one spouse was needed at home, it was the highest-paid spouse who stayed in the paid workforce. Shift work, especially double shifts, dramatically increased the wages of steelworkers.

16. Byington, *Homestead*, 37, 171–72.

17. Byington, *Homestead*, 37, 171–72.

18. Fitch, *The Steel Workers*, 62–63.

19. Byington, *Homestead*, 64. See also Modell, *A Town Without Steel*, 94.

20. For other examples of breadwinners in the 1950s who resented wives being involved in activities that distracted them from giving attention to their husbands, see Coontz, *The Way We Never Were*, 165; and Mercier, *Anaconda*, 127.

21. For a discussion of another suburban industrial community in the 1950s, see Bennet M. Berger, *Working-Class Suburb: A Study of Auto Workers in Suburbia* (Berkeley and Los Angeles: University of California Press, 1968), especially chapter 5.

22. John Bodnar, *Workers' World: Kinship, Community, and Protest in an Industrial Society, 1900–1940* (Baltimore: Johns Hopkins University Press, 1982), 13; Bruno, *Steelworker Alley*, 48–49; Micaela di Leonardo, "The Female World of Cards and Holidays: Women, Families, and the Work of Kinship," *Signs: Journal of Women in Culture and Society* 12, no. 3 (1987): 246–61; Barbara Wernecke Durkin, *Oh, You Dundalk Girls, Can't You Dance the Polka?* (New York: William Morrow, 1984); Mintz and Kellogg, *Domestic Revolutions*, 94–95.

23. Historians of family violence argue that drinking is another way that men claim dominance within their families, but that drinking is not the cause of family violence. Until recently, the wives of steelworkers excused their husbands' drinking because "after working that hard men need to drink." Halle, *America's Working Man*, 62–64; Mercier, *Anaconda*, 29; Linda Gordon, *Heroes of Their Own Lives: The Politics and History of Family Violence* (New York: Viking Penguin, 1988).

24. Byington, *Homestead*, 22–40; Diggs, *From the Meadows to the Point*, 111–15; Fitch, *The Steel Workers*, 9–11; Modell, *A Town Without Steel*, 93; Pappas, *The Magic City*, 182–83.

25. Coontz, *The Way We Never Were*, 30; Matthews, *"Just a Housewife", 210.*

26. Halle, *America's Working Man*, 66; Cohen, *A Consumers' Republic*, 173.

27. For an explanation of the ways in which wives and their husbands collaborated in breadwinner/homemaker family dynamics, see Halle, *America's Working Man*, 66.

28. Bruno, *Steelworker Alley*, 94–96; Hartmann, "Women's Employment and the Domestic Ideal in the Early Cold War Years," 84–86; Modell, *A Town Without Steel*, 100. The devastating effects of layoffs on blue-collar workers is described in Bruno, *Steelworker Alley*, 84–89; and Pappas, *The Magic City*, 80–86.

29. Coontz, *The Way We Never Were*, 32; Modell, *A Town Without Steel*, 10, 100. Coontz cites an article from a 1954 edition of *Esquire* that described working wives as a "menace."

30. Bruno, *Steelworker Alley*, 84–89; Pappas, *The Magic City*, 80–86.

31. Matthews, *"Just a Housewife"*, 208.

32. Byington, *Homestead*, 154; Halle, *America's Working Man*, 64–67. Byington reports that steelworkers gave their wages to their wives, who in turn did the budgeting and paid the bills. Halle indicates that by the 1950s, male steelworkers controlled the family budgeting and spending.

33. Wini Breines, *Young, White, and Miserable: Growing Up Female in the Fifties* (Boston: Beacon Press, 1992); Joan Williams, *Unbending Gender: Why Family and Work Conflict and What To Do About It* (New York: Oxford University Press, 2000), 157–60; "How America Lives," *Ladies Home Journal*, April 1945.

34. Susan Hartmann, *The Home Front and Beyond: American Women in the 1940s* (Boston: Twayne, 1982), 84–100.

CHAPTER 5

1. See Ava Baron, "Gender and Labor History: Learning from the Past, Looking to the Future," in Baron, *Work Engendered*, 1–46; Brody, *Steelworkers in America*; Byington, *Homestead*, 32–36; Fitch, *The Steel Workers*; and Reutter, *Sparrows Point*.

2. Kay Deaux and Joseph C. Ullman, *Women of Steel: Female Blue-Collar Workers in the Basic Steel Industry* (New York: Praeger Publishers, 1983); Mary Margaret Fonow, *Union Women: Forging Feminism in the United Steelworkers of America* (Minneapolis: University of Minnesota Press, 2003); Reutter, *Sparrows Point*, 157–59.

3. See Hine and Thompson, *A Shining Thread of Hope*; Barbara Klaczynska, "Why Women Work," *Labor History* 17 (Winter 1976): 73–87.

4. See Reutter, *Sparrows Point*, 360. Also see Baron, *Work Engendered*, 179.

5. Reutter, *Sparrows Point*, 360–67.

6. Reutter, *Sparrows Point*, 360–67.

7. Joy Parr, *The Gender of Breadwinners: Women, Men, and Change in Two Industrial Towns, 1880–1950* (Toronto: University of Toronto Press, 1990), 104.

8. Peiss, *Hope in a Jar*, 145; Reutter, *Sparrows Point*, 364–65.

9. Reutter, *Sparrows Point*, 371–73.

10. Peiss, *Hope in a Jar*, 173–74.

11. See Sherna Berger Gluck, *Rosie the Riveter Revisited: Women, the War, and Social Change* (New York: Twayne, 1987); Karen Anderson, *Wartime Women: Sex Roles, Family Relations, and the Status of Women During World War II* (Westport, Conn.: Greenwood Press, 1981); Jamie Stiehm, "'Rosies' Share Stories of Their Riveting Work," *Baltimore Sun*, June 13, 2004, 1B, 3B; Darlene Clark Hine, "Mabell K. Staupers and the Integration of Black Nurses into the Armed Forces During World War II," in Hine, *Hine Sight*, 183–201; Maureen Honey, *Creating Rosie the Riveter: Class, Gender, and Propaganda During World War II* (Amherst: University of Massachusetts Press, 1984); Terri Narrell Mause, "Rosie the Riveter To Be Honored During DCC Program Next Week," *Dundalk Eagle*, March 9, 1995, 1, 18; James D. Rose, "'The Problem Every Supervisor Dreads': Women Workers at the U.S. Steel Duquesne Works During World War II," *Labor History* 36 (Winter 1995): 24–51; and Skold, "The Job He Left Behind." Also see the government publications that give statistics on women war workers during World War II, including *Baltimore Women War Workers in the Postwar Period* (Washington, D.C.: Government Printing Office, 1948) and *Women's Work in the War* (Washington, D.C.: Government Printing Office, 1942).

12. For a description of women in the steel industry during World War II, see Ethel Erickson, "Women's Employment in the Making of Steel, 1943," Bulletin of the Women's Bureau, no. 192–95 (Washington, D.C.: Government Printing Office, 1944).

13. Karen Anderson, "Last Hired, First Fired: Black Women Workers During World War II," *Journal of American History* 69 (June 1982): 82–97; Diggs, *From the Meadows to the Point*; Melissa Dabakis, "Norman Rockwell's *Rosie the Riveter* and the Discourses of Wartime Womanhood," in Melosh, *Gender and American History Since 1890*, 182–204; Milkman, *Gender at Work*, 66–101.

14. Anderson, *Wartime Women*, 43–46; Dabakis, "Norman Rockwell's *Rosie the Riveter* and the Discourses of Wartime Womanhood"; Gluck, *Rosie the Riveter Revisited*; Brigid O'Farrell and Suzanne Moore, "Unions, Hard Hats, and Women Workers," in *Women and Unions: Forging a Partnership*, ed. Dorothy Sue Cobble (Ithaca: ILR Press, 1993), 79–82; Modell, *A Town Without Steel*, 67.

15. O'Farrell and Moore, "Unions, Hard Hats, and Women Workers," 80–83.

16. Anderson; Gluck, *Rosie the Riveter Revisited*, 37–39; Dukes, "Remembering What's Important," in *The Heat*, 25; and Halle, *America's Working Man*, 56.

17. Karen Anderson argues that women were happy with nontraditional jobs. Sherna Berger Gluck points out that some women had to work and others chose to work; some were

glad to go back home and others wanted to stay employed. See Gluck, *Rosie the Riveter Revisited*, 52–53, 63, 170.

18. Statistics on union membership are from the *Baltimore Sun* and the USWA. For a discussion of women's roles in other male-dominated unions, see Cobble, *Women and Unions*; Nancy F. Gabin, *Feminism in the Labor Movement: Women and the United Auto Workers, 1935–1975* (Ithaca: Cornell University Press, 1990); Alice Kessler-Harris, "Problems of Coalition-Building: Women and Trade Unions in the 1920s," in Ruth Milkman, ed., *Women, Work, and Protest: A Century of Women's Labor History*, 110–38 (Boston: Routledge, 1985). In his 1886 address, "Labor Union's Address to the Workingmen of the United States," Thomas A. Armstrong emphasized the union position that there be unequal pay for women. Krause, *The Battle for Homestead, 1880–1892*, 96, 124, 206.

19. *Steel Labor: Voice of the United Steelworkers of America* 1, no. 1 (August 1936) through 44, no. 11 (November 1979). Also see Gabin, *Feminism in the Labor Movement*; George Lipsitz, *Rainbow at Midnight: Labor and Culture in the 1940s* (Urbana: University of Illinois Press, 1994); and Modell, *A Town Without Steel*, 164.

20. For a comparison with the inclusion of women autoworkers in the United Auto Workers union, see Gabin, *Feminism in the Labor Movement*, 229–36.

21. For a discussion of the significantly changed relationship between women and unions, see Dorothy Sue Cobble, "Rethinking Troubled Relations Between Women and Unions: Craft Unionism and Female Activism," *Feminist Studies* 16 (Fall 1990): 519–48. Ruth Milkman has written extensively on the changing relationship of American unions to the growing number of women in the paid workforce. Industrial unions like the USWA were typical of blue-collar unions with a long history of serving a male constituency, a situation that ensures, according to Milkman, that the "dominant cultural imagery of labor organization and of union power remains male." Ruth Milkman, "Women Workers, Feminism and the Labor Movement since the 1960s," in Milkman, *Women, Work, and Protest*, 306. Also see Dukes, "The Card," in *The Heat*, 36–38.

22. Halle, *America's Working Man*, 55; Skold, "The Job He Left Behind," 60–61.

23. Bensman and Lynch, *Rusted Dreams*, 90–96; Gluck, *Rosie the Riveter Revisited*, 85; Barbara F. Reskin and Patricia A. Roos, *Job Queues, Gender Queues: Explaining Women's Inroads into Male Occupations* (Philadelphia: Temple University Press, 1990), 135–42; Robinson, *Affirmative Action*, 185–88; Skold, "The Job He Left Behind," 66–68.

24. Dukes, "Remembering What's Important," in *The Heat*, 21–30; Halle, *America's Working Man*, 44–46.

25. Bensman and Lynch, *Rusted Dreams*, 170; Gluck, *Rosie the Riveter Revisited*, 52–53; Jones, "Day One," in *The Heat*, 27–30; Mintz and Kellogg, *Domestic Revolutions*, 207.

26. Cobble, *Women and Unions*, 55–78; Baron, *Work Engendered*, 209; Alice H. Cook, Val R. Lorwin, and Arlene Kaplan Daniels, *The Most Difficult Revolution: Women and Trade Unions* (Ithaca: Cornell University Press, 1992); Margaret Hallock, "Unions and the Gender Wage Gap," in *Women and Unions: Forging a Partnership*, ed. Dorothy Sue Cobble (Ithaca: Cornell University Press, 1993), 27–42.

27. Jennifer Jones, "Women in the Mill," in *The Heat*, 129; Martha May, "Bread Before Roses: American Workingmen, Labor Unions and the Family Wage," in Ruth Milkman, ed., *Women, Work, and Protest* (New York: Routledge, 1985), 1–21.

28. Bensman and Kellogg, *Rusted Dreams*, 26; Dukes, "Remembering What's Important," in *The Heat*, 23; Mary Margaret Fonow, "Occupation/Steelworker: Sex/Female," in *Feminist Frontiers*, ed. Laurel Richardson and Verta Taylor, 217–222; Christine L. Williams, *Gender Differences at Work: Women and Men in Nontraditional Occupations* (Berkeley and Los Angeles: University of California Press, 1989).

29. Modell, *A Town Without Steel*, 42; Skolnick, *Embattled Paradise*, 107; Rosalyn Terborg-Penn, "African American Women and the Vote: An Overview," in *African American Women and the Vote, 1837–1965*, ed. Ann D. Gordon, Bettye Collier-Thomas, John H. Bracey, Arlene Voski Avakian, and Joyce Avrech Berkman (Amherst: University of Massachusetts Press, 1997), 10–23.

30. This range of responses to women who began doing production work in traditionally male industries in the 1970s has been noted by other researchers: "blue-collar male attitudes are diverse, complex, and sometimes surprising. . . . Men seem to fit into three categories: a small group of actively supportive men, a minority of very hostile men, and the majority, who were neutral or ambivalent." O'Farrell and Moore, "Unions, Hard Hats, and Women Workers," 79. See also Bruno, *Steelworker Alley*, 71–72.

31. Gluck, *Rosie the Riveter Revisited*, 190.

32. Halle, *America's Working Man*, 57; Mercier, *Anaconda*, 71–72.

33. Bensman and Lynch, *Rusted Dreams*, 26; Halle, *America's Working Man*, 62–64; Jones, "Women in the Mill," in *The Heat*, 129; Skolnick, *Embattled Paradise*, 113; Clara Bingham and Laura Leedy Gansler, *Class Action: The Story of Lois Jenson and the Landmark Case that Changed Sexual Harassment Law* (New York: Doubleday, 2002); Barbara Kantrowitz, "Striking a Nerve," *Newsweek*, October 21, 1991: 34–40, as quoted in Baxandall and Gordon, *America's Working Women*; Kerry Segrave, *The Sexual Harassment of Women in the Workplace, 1600–1993* (Jefferson, N.C.: McFarland and Company, 1994).

34. Halle, *America's Working Man*, 56.

35. Heidi Hartmann, "Roundtable on Affirmative Action and Pay Equity," in Cobble, *Women and Unions*, 43–49; Lois S. Gray, "The Route to the Top: Female Union Leaders and Union Police," in Baron, *Work Engendered*, 378–93; Dukes, "The Card," in *The Heat*, 36–38; Jones, "Women in the Mill," in *The Heat*, 129; Modell, *A Town Without Steel*, 185.

36. Working in a steel mill is still unappealing to many women and to many men. Several steelworkers said that they hoped their sons would not end up in the mill; none of them wanted their daughters working in steel. Christine L. Williams, *Still a Man's World* (Berkeley and Los Angeles: University of California Press, 1995); Halle, *America's Working Man*, 99–102; Woodring, "Men of Steel, Hearts of Gold," in *The Heat*, 96.

CHAPTER 6

1. For discussions of rust belt communities that had been entirely dependent on the steel mills that closed in the 1980s, see Bensman and Lynch, *Rusted Dreams*; Bruno, *Steelworker Alley*; Donald F. Barnett and Robert W. Crandall, *Up from the Ashes: The Rise of the Steel Minimill in the United States* (Washington, D.C.: The Brookings Institute, 1986); John Hoerr, *And the Wolf Finally Came: The Decline of the American Steel Industry* (Pittsburgh: University of Pittsburgh Press, 1988); Lynd, *The Fight Against Shutdowns*; Modell, *A Town Without Steel*; Terry F. Buss and F. Stevens Redburn, *Reemployment After a Shutdown: The Youngstown Steel Mill Closings, 1977–1985* (Youngstown: Center for Urban Studies, Youngstown State University, 1986); Robert W. Crandall, *The U.S. Steel Industry in Recurrent Crisis: Policy Options in a Competitive World* (Washington, D.C.: Brookings Institution, 1981); Tom Juravich and Kate Bronfenbrenner, *Ravenswood: The Steelworkers' Victory and the Revival of American Labor* (Ithaca: Cornell University Press, 1999); Richard Preston, *American Steel: Hot Metal Men and the Resurrection of the Rust Belt* (New York: Prentice Hall, 1991); William Serrin, *Homestead: The Glory and Tragedy of an American Steel Town* (New York: Times Books, 1992); John Strohmeyer, *Crisis in Bethlehem: Big Steel's Struggle to Survive* (Bethesda, Md.: Adler and Adler, 1986); and Paul A. Tiffany, *The Decline of American Steel: How Management, Labor and Government Went Wrong* (New York: Oxford University Press, 1988).

2. Statistics on union membership come from the *Baltimore Sun* and the USWA Local 2609 and Local 2610. See also Marty Marciniak, "Cobbled and Hobbled," in *The Heat*, 137–38; and Eileen Appelbaum and Rosemary Batt, *The New American Workplace* (Ithaca: Cornell University Press, 1994).

3. Michael Hill, "Down and Out in Dundalk," *Baltimore Sun*, December 27, 1992; Milt Schwartzman, "Dundalk's Spirit," *Baltimore Sun*, January 5, 1993.

4. Byington, *Homestead*; Diggs, *From the Meadows to the Point*, 44–46; Fitch, *The Steel Workers*; Modell, *A Town Without Steel*, 93; Dundalk-Patapsco Neck Historical Society, "Re-

flections"; Daniel T. Rogers, *The Work Ethic in Industrial America, 1850–1920* (Urbana: University of Illinois Press, 1978), 102.

5. Modell, *A Town Without Steel*, 105–7; Pappas, *The Magic City*, 30–31.

6. *Directory of Major Manufacturers* (Baltimore: Greater Baltimore Committee, 1981).

7. *Directory of Major Manufacturers.*

8. Baltimore Metropolitan Planning Council, *Report of the Baltimore Metropolitan Planning Council* (Baltimore: Author, March 1982).

9. Baltimore Metropolitan Planning Council, *Report of the Baltimore Metropolitan Planning Council.*

10. Baltimore Metropolitan Planning Council, *Report of the Baltimore Metropolitan Planning Council.*

11. For a discussion of women's commitment to long-term employment during the 1950s, see Weiner, *From Working Girl to Working Mother*, 89–96

12. For a historical perspective on men as providers and breadwinners, see Julie Matthaei, *An Economic History of Women in America* (New York: Schocken Books, 1982), 211–18.

13. The varying responses to single-industry communities losing their source of abundant unionized jobs is discussed in Barry Bluestone and Bennett Harrison, *The Deindustrialization of America: Plant Closings, Community Abandonment, and the Dismantling of Business and Industry* (New York: Basic Books, 1982); Bruno, *Steelworker Alley*; Mercier, *Anaconda*; Dale Russakoff, "Chasing Work Fractures Lives of Steel Families," *Washington Post*, March 28, 1993, A1, A8; and Robert Ward, *Red Baker* (Garden City, N.Y.: Doubleday, 1985).

14. Jessie Bernard, "The Good-Provider Role: Its Rise and Fall," *American Psychologist* 36, no. 1 (1981): 1–12. On the loss of the breadwinner/provider role, see Bruno, *Steelworker Alley*, 149–53; Doc Iler, "A Violet in the Light," in *The Heat*, 154–57; Mirra Komarovsky and Michael S. Kimmel, *The Unemployed Man and His Family: The Effect of Unemployment Upon the Status of the Man in Fifty-nine Families* (1927; reprint, New York: Altamira Press, 2004), 32–57; Melosh, *Gender and American History Since 1890*, 181; Mercier, *Anaconda*, 123; Modell, *A Town Without Steel*, 43; Pappas, *The Magic City*, 80–86; Weiner, *From Working Girl to Working Mother*, 104.

15. George F. Becker, "Preface," in *The Heat*, 7–8; Pappas, *The Magic City*, 183–85; Annette Bernhardt, Martina Morris, and Mark Hancock, "Women's Gains or Men's Losses? A Closer Look at the Shrinking Gender Gap in Earnings," *American Journal of Sociology* 101, no. 2 (1995): 302–28; Barbara Ehrenreich, *Nickled and Dimed: On (Not) Getting by in America* (New York: Henry Holt, 2001); Richard Freeman, "How Much Has De-Unionization Contributed to the Rise in Male Earning Inequality?" in *Uneven Tides: Rising Inequality in America*, ed. Shelton Danziger and Peter Gottschalk, 100-121 (New York: Russell Sage Foundation, 1993).

16. Many steelworkers, and even union officials of USWA Local 2609 and Local 2610, reported that they ultimately were grateful to get out of unhealthy jobs. Joseph Pleck, *Working Wives, Working Husbands* (Beverly Hills, Calif.: Sage, 1985); Jean L. Potuchek, *Who Supports the Family? Gender and Breadwinning in Dual-Earner Marriages* (Stanford: Stanford University Press, 1997).

17. Susan Porter Benson, "Living on the Margin: Working-Class Marriages and Family Survival Strategies in the United States," in *The Sex of Things: Gender and Consumption in Historical Perspective*, ed. Victoria de Grazia (Berkeley and Los Angeles: University of California Press, 1996), 212; Mirra Komarovsky, *Blue-Collar Marriage* (New York: Random House, 1962), 133; Mercier, *Anaconda*, 7, 203–4.

18. Halle, *America's Working Man*, 299; Komarovsky, *Blue-Collar Marriage*, 42–53.

19. Landry, *Black Working Wives*; Hidreth Y. Grossman and Nia Lane Chester, eds., *The Experience and Meaning of Work in Women's Lives* (Hillsdale, N.J.: Lawrence Erlbaum Associates, 1990); Tamar Lewin, "Women Are Becoming Equal Providers," *New York Times*, May 11, 1995, A27; Mercier, *Anaconda*, 90, 203–4; Mintz and Kellogg, *Domestic Revolutions*, 204–9; Skolnick, *Embattled Paradise*, 107–9.

20. See Louise Lamphere, *From Working Daughters to Working Mothers: Immigrant Women in a New England Industrial Community* (Ithaca: Cornell University Press, 1987), 269–70. Lamphere discusses the ways in which the husband's role as provider is "bound up with manly honor." Like the Columbian and Portuguese men that Lamphere studied, Dundalk men may continue to cherish the ideal of being the breadwinner of their families, but they are willing to adjust that ideal to meet the new financial situations that face their families. See also Jane Wheelock, *Husbands at Home: The Domestic Economy in a Post-Industrial Society* (New York: Routledge, 1990).

21. Komarovsky, *Blue-Collar Marriage*, 71–83.

22. See Weiner, *From Working Girl to Working Mother*, 3–9.

23. Bluestone and Harrison, *The Deindustrialization of America.*

24. Bodnar, *Steelton*; Byington, *Homestead*, 57; Modell, *A Town Without Steel*, 91.

25. For a discussion of the shift in consensus about working wives, see Weiner, *From Working Girl to Working Mother*, 139–40.

26. Mintz and Kellogg, *Domestic Revolutions*, 198–99; Wendy Luttrell, "'The Teachers, They All Had Their Pets': Concepts of Gender, Knowledge, and Power," *Signs: Journal of Women in Culture and Society* 18 (Spring 1993): 505–46.

27. Coontz, *The Way We Never Were*, 161–68. makes the point that women's ideological position often changed after they were forced to go to work.

28. Jennifer Johnson, *Getting By On the Minimum: The Lives of Working-Class Women* (New York: Routledge, 2002), 103.

29. Johnson, *Getting By On the Minimum*, 102; Mercier, *Anaconda*, 75; Skolnick, *Embattled Paradise*, 107–9.

30. Weiner, *From Working Girl to Working Mother*, 105.

31. Gluck, *Rosie the Riveter Revisited*, 84; Johnson, *Getting By On the Minimum*, 103; Grossman and Chester, *The Experience and Meaning of Work in Women's Lives.*

32. See Myra Marx Ferree, "Working-Class Jobs: Paid Work and Housework as Sources of Satisfaction," *Social Problems* 23 (April 1976): 431–41.

33. See Ellen Israel Rosen, *Bitter Choices: Blue-Collar Women In and Out of Work* (Chicago: University of Chicago Press, 1987).

CHAPTER 7

1. Lipsitz, *Rainbow at Midnight*, 55; Oppenheimer, *The Female Labor Force in the United States*, 188–89; Tentler, *Wage-Earning Women*, 136–79; Alice Kessler-Harris, *A Woman's Wage: Historical Meanings and Social Consequences* (Lexington: University Press of Kentucky, 1990).

2. For discussion of men giving up the breadwinner role, see Cobble, *Women and Unions*; May, "Bread Before Roses," 3, 19; Mercier, *Anaconda*, 49, 188.

3. Eventually, media such as *Martha Stewart Living* would turn shopping into a well-crafted art for women. See Cohen, *A Consumers' Republic*, 36–39, 316–18; Rosemary Marangoly George, ed., *Burning Down the House: Recycling Domesticity* (Boulder, Colo.: Westview Press, 1998); Matthews, *"Just a Housewife"*, 212; Peiss, *Hope in a Jar*, 173–74.

4. Cohen, *A Consumers' Republic*, 189.

5. Halle, *America's Working Man*, 62; Johnson, *Getting By On the Minimum*, 102; Skolnick, *Embattled Paradise*, 107–9.

6. Weiner, *From Working Girl to Working Mother*, 104–5.

7. For the best study of contemporary two-job families, see Hochschild, *The Second Shift*. Hochschild distinguishes between the traditional ideal and the egalitarian ideal for sharing responsibilities within a marriage. She then points out that marriages are not categorically one or the other:

> Indeed, a split between these two ideals seemed to run not only between social classes, but between partners within marriages and between two contending voices inside the conscience of one individual. The working classes tended toward the traditional ideal, and the middle class tended toward the egalitarian. Men tended toward the traditional idea, women toward the egalitarian one. . . . Most marriages were either torn by, or a settled compromise between, these two ideals. In this sense, the split between them runs implicitly through every marriage I came to know. (189)

See also Karen D. Fox and Sharon Y. Nickols, "The Time Crunch: Wife's Employment and Family Work," *Journal of Family Issues* 4 (March 1983): 61–82; Jacqueline J. Goodnow and Jennifer M. Boeves, *Men, Women and Household Work* (New York: Oxford University Press, 1994); Arlie Russell Hochschild, *The Time Bind: When Work Becomes Home and Home Becomes Work* (New York: Henry Holt, 1997); Bart Landry, *Black Working Wives: Pioneers of the American Family Revolution* (Berkeley and Los Angeles: University of California Press, 2000). Tamar Lewin, "Men Assuming Bigger Share At Home, New Survey Shows," *New York Times*, April 14, 1998; Elizabeth Pleck, "Two Worlds in One: Work and Family," *Journal of Social History* 10, no. 2 (1976): 178–95. It was not uncommon for studies of shared housework in dual career households to ignore African American families, but this group is included in Hochschild, *The Second Shift*, 129–41; Lillian Rubin, *Families on the Fault Line* (New York: HarperCollins, 1994), 104; Joan Williams, 154–55.

8. Glenda Riley, *Divorce: An American Tradition* (New York: Oxford University Press, 1991), 186; and Skolnick, *Embattled Paradise*, 112.

9. Cohen, *A Consumers' Republic*, 148.

10. Skolnick, *Embattled Paradise*, 208.

11. Sara Evans, *Tidal Wave: How Women Changed America at Century's End* (New York: Free Press, 2003), 167.

12. Cohen, *A Consumers' Republic*, 485.

13. Susan J. Douglas, *Where the Girls Are: Growing Up Female with the Mass Media* (New York: Three Rivers Press, 1994), 79; Evans, *Tidal Wave*, 85–89, 110; Mintz and Kellogg, *Domestic Revolutions*, 208. The common disclaimer by contemporary women—"I'm not a feminist but . . ."—is discussed in Douglas, *Where the Girls Are*, 278–94.

14. Mintz and Kellogg, *Domestic Revolutions*, 208.

15. Evans, *Tidal Wave*, 139; Modell, *A Town Without Steel*, 50.

16. Douglas, *Where the Girls Are*, 278–79; Halle, *America's Working Man*, 55–56.

17. Mintz and Kellogg, *Domestic Revolutions*, 274 n. 23.

18. Halle, *America's Working Man*, 55–56; Williams, *Unbending Gender*, 147–50.

19. See Lamphere, *From Working Daughters to Working Mothers*, 269. Lamphere discusses the ways in which employment brings women into contact with a network of coworkers that replaces the network of friends, neighbors, and kin comprising the social world of the homemaker.

20. Sarah F. Berk, *The Gender Factory: The Apportionment of Work in American Households* (New York: Plenum Press, 1985); Heidi Hartmann, "The Family as the Locus of Gender, Class and Political Struggle: The Example of Housework," *Signs: Journal of Women in Culture and Society* 6 (Spring 1981): 366–94; Mary Claire Lennon and Sarah Rosenfield, "Relative Fairness and the Division of Housework: The Importance of Options," *American Journal of Sociology* 100, no. 2 (1994): 506–31; Mintz and Kellogg, *Domestic Revolutions*, 274; Parr, *The Gender of Breadwinners*, 301.

21. Mintz and Kellogg, *Domestic Revolutions*, 295 n. 22; Williams, *Unbending Gender*, 55–56.

22. For a statistical analysis of how much housework is actually shared by husbands and wives, see Hochschild, *The Second Shift*, 216–38; Williams, *Unbending Gender*, 2.

23. Despite social sanctions against divorce and against married women working out-

side the home, there was a significant increase in both during the 1950s. See Coontz, *The Way We Never Were*; and Riley, *Divorce*, 156.

24. Mintz and Kellogg, *Domestic Revolutions*, 128–29; Williams, *Unbending Gender*, 146.

25. Coontz, *The Way We Never Were*, 161–68; Ogden, *The Great American Housewife*, 196; Riley, *Divorce*, 88–89.

26. Victor R. Fuchs, *How We Live* (Cambridge: Harvard University Press, 1983), 147–50; Halle, *America's Working Man*, 72–73; Mintz and Kellogg, *Domestic Revolutions*, 203–7.

27. Francis Ivan Nye and Lois Wlapis Hoffman, eds., *The Employed Mother in America* (Chicago: Rand McNally, 1963), 37–53. Pressures on women to excel in both the paid workforce and with child care are discussed in Williams, *Unbending Gender*, 58–59, 146–53.

CHAPTER 8

1. Jo Ann E. Argersinger, *Toward a New Deal in Baltimore: People and Government in the Great Depression* (Chapel Hill: University of North Carolina Press, 1988), 1–20; Robert D. Bullard, J. Eugene Gribsby II, and Charles Lee, eds., *Residential Apartheid: An American Legacy* (Berkeley and Los Angeles: University of California Press, 1994), 82–98; Diggs, *From the Meadows to the Point*, 214; Dundalk-Patapsco Neck Historical Society, "Reflections."

2. Willie, *Black and White Families*, 270–89; Hariette McAdoo, *Black Families* (Thousand Oaks, Calif.: Russell Sage Foundation, 1997).

3. Byington, *Homestead*; Dickerson, *Out of the Crucible*; Nelson, *Divided We Stand*; Modell, *A Town Without Steel*. The selective process of memories of experiences shared by blacks and whites is addressed in Blight, *Race and Reunion*; and Shackel, *Memory in Black and White*.

4. Chafe, Gavins, and Korstad, *Remembering Jim Crow*; Clark-Lewis, *Living In, Living Out*; Phyllis Palmer, *Domesticity and Dirt: Housewives and Domestic Servants in the United States, 1920–1945* (Philadelphia: Temple University Press, 1989).

5. Lynetric Bridges, "Domestic Workers in Baltimore, 1880–1975" (master's thesis, Morgan University, 2002); Clark-Lewis, *Living In, Living Out*, 82–90; and Hine and Thompson. *A Shining Thread of Hope*; Vron Ware and Les Back, *Out of Whiteness: Color, Politics, and Culture* (Chicago: University of Chicago Press, 2002).

6. Anderson, *Wartime Women*; Mause, "Dundalk Timeline," *Dundalk Eagle*, n.d.; P. David Woodring, "Men of Steel, Hearts of Gold," in *The Heat*, 90–96; Maryland Public Television, *A Hometown at War*.

7. Bullard, Gribsby, and Lee, *Residential Apartheid*, 82–98; Diggs, *From the Meadows to the Point*, 87; W. Dennis Keating, *The Suburban Racial Dilemma: Housing and Neighborhoods* (Philadelphia: Temple University Press, 1994), 473; Dundalk-Patapsco Neck Historical Society, "Reflections," iii, 38, 47.

8. Chafe, Gavins, and Korstad, *Remembering Jim Crow*, 124–28, 152–204; Diggs, *From the Meadows to the Point*, 76, 217; Gwendolyn Etter-Lewis, *My Soul Is My Own: Oral Narratives of African American Women in the Professions* (New York: Routledge, 1993), 92–104; Ted Patterson, "The Educational Journey of African Americans in Southeastern Baltimore County" (unpublished manuscript, n.d.); Shaw, *What A Woman Ought To Be and To Do*, 33–40; Dail Willis, "Sollers Point High's First Graduates Return: Despite Racial Barriers, Class of '49 Perseveres," *Baltimore Sun*, April 25, 1999.

9. Cowan, *More Work for Mother*, 120; Diggs, *From the Meadows to the Point*, 50, 79, 104–9.

10. African American women have always served as household heads at a higher rate than have white women, with or without Aid to Families with Dependent Children. See Carole Shammas, *A History of Household Government in America* (Charlottesville: University of Virginia Press, 2002); Donna Franklin, *Ensuring Inequality: The Structural Transformation of the African American Family* (New York: Routledge, 1997); Steven Ruggles, "The Origins of African-American Family Structure," *American Sociological Review* 59 (February 1994): 136–51.

11. Hine and Thompson, *A Shining Thread of Hope*, 226; Shaw, *What a Woman Ought To Be and To Do*, 42–45. The explicit articulation of social and political consciousness and activism on the part of African American women is discussed in Patricia Hill Collins, *Black Feminist Thought: Knowledge, Consciousness, and the Politics of Empowerment* (London: Harper-Collins Academic, 1990); and Gordon, Collier-Thomas, Bracey, Avakian, and Berkman, eds., *African American Women and the Vote*, 10–23.

12. The focus of the discussion of gender and race relations in southeast Baltimore County at the turn of the twenty-first century relies on the insights of cultural geography, including Susan Hanson and Geraldine Pratt, *Gender, Work, and Space* (New York: Routledge, 1995); Doreen Massey, *Politics and Method: Contrasting Studies In Industrial Geography* (London: Macmillan, 1984); Mark B. Miller, *Baltimore Transitions: Views of an American City in Flux*, rev. ed. (Baltimore: Johns Hopkins University Press, 1999).

13. Hine, *Hine Sight*, 225, 263; Mercier, *Anaconda*, 121–22; Modell, *A Town Without Steel*, 30; Shaw, *What a Woman Ought To Be and To Do*, 38–40.

14. This is mentioned in other accounts of women leaving self-contained industrial communities for more diverse metropolitan areas. See Teresa Amott, *Caught in the Crisis: Women and the U.S. Economy Today* (New York: Monthly Review Press, 1993); Hanson and Pratt, *Gender, Work, and Space*. The movement of African American women into an array of blue-collar and professional employment niches is palpable in the Baltimore metropolitan area. National studies on this phenomenon include Cunningham and Zalozar, "The Economic Progress of Black Women, 1940–1980;" Sokoloff, *Black Women and White Women in the Professions*; Jane Riblett Wilkie, "The Decline of Occupational Segregation Between Black and White Women," in *Research in Race and Ethnic Relations: A Research Annual*, ed. Cora Bagley Marrett and Cheryl Leggon, 4 (Fairfax, Va.: JIA Publisher, 1982), 84–97.

15. Michelle Tea, ed., *Without A Net: The Female Experience of Growing Up Working Class* (Emeryville, Calif.: Seal Press, 2003); and Willie, *Black and White Families*, 19. Expressions of class consciousness and class interests have been altered in southeast Baltimore County with the decline of the union movement led by the USWA in the Dundalk and Turner Station communities. See Joseph H. Boyett and H. P. Conn, *Workplace 2000: The Revolution Reshaping America* (New York: Dutton, 1992); John R. Hall, ed., *Reworking Class* (Ithaca: Cornell University Press, 1997); John Hinshaw and Paul Le Blanc, eds., *U.S. Labor in the Twentieth Century: Studies in Working-Class Struggles and Insurgency* (Amherst, N.Y.: Humanity Press, 2000); and David Harvey, *Consciousness and the Urban Experience* (Baltimore: Johns Hopkins University Press, 1985).

16. McCall, *Complex Inequality*, 146–45, 155–68; Peiss, *Hope in a Jar*, 145–46; Reeve Vanneman and Lynn Weber Cannon, *The American Perception of Class* (Philadelphia: Temple University Press, 1987).

17. Richard Sennett and Jonathan Cobb, *The Hidden Injuries of Class* (New York: W. W. Norton, 1972); Tea, *Without a Net*; Chris Kaltenbach, "Brian Wilson Was Just Being Himself: The Personality That Made the Dundalk Bashing DJ Famous Is the Same One That Got Him Fired," *Baltimore Sun*, November 26, 1995.

18. Sennett and Cobb, *The Hidden Injuries of Class*, 16–27; Richard Rodriguez, *The Ironies of Education* (New York: Routledge, 1999).

19. Halle, *America's Working Man*, 55–59; Sennett and Cobb, *The Hidden Injuries of Class*, 52–67; Ira Katznelson, *City Trenches: Urban Politics and the Patterning of Class in the United States*.

20. Hall, *Reworking Class*; Alfred Lubrano, *Limbo: Blue-Collar Roots, White-Collar Dreams* (Waltham, Mass.: Reed Business Information, 2003); Beverly Skeggs, *Class, Self, Culture* (New York: Routledge, 2004).

21. Hall, *Reworking Class*, 87–79; Gareth Stedman Jones, *Languages of Class* (New York: Cambridge University Press, 1983); Pooja Makhijani, *How Girls Experience Race in America* (Emeryville, Calif.: Seal Press, 2004).

22. Frank Levy, *Dollars and Dreams: The Changing American Income Distribution* (New

York: Norton and Russell Sage Foundation, 1988); Boyett and Conn, *Workplace 2000*; Halle, *America's Working Man*, 48.

23. Rodriguez, *The Ironies of Education.*

24. C. P. Ellis, quoted in Studs Terkel, *American Dreams: Lost and Found* (New York: Ballantine, 1987); Derek Bok, *Beyond the Ivory Tower: Social Responsibilities of the Modern University* (Cambridge: Harvard University Press, 1982).

25. David Truscello, "Expanding Social Capital Networks of Knowledge: A Critical Pedagogy" (Ph.D. diss., University of Maryland Baltimore County, 2004); Vanneman and Cannon, *The American Perception of Class.*

26. Dundalk-Patapsco Neck Historical Society, "Reflections," 33.

27. Halle, *America's Working Man*, 90–91.

28. Halle, *America's Working Man*, 88–89.

29. U.S. Census Bureau, Twenty-second Census, 2000. The statistics from the 2000 census indicate that Turner Station will become older and poorer in the next decade. For discussions of contemporary interracial social discourse, see David L. Kirp, John P. Dwyer, and Larry A. Rosenthal, *Our Town: Race, Housing, and the Soul of Suburbia* (New Brunswick: Rutgers University Press, 1995); Harold A. McDougall, *Black Baltimore: A New Theory of Community* (Philadelphia: Temple University Press, 1993); and W. Edward Orser, *Blockbusting in Baltimore: The Edmondson Village Story* (Lexington: University of Kentucky Press, 1994).

30. Diggs, *From the Meadows to the Point*, 53.

31. Diggs, *From the Meadows to the Point*, 79; Halle, *America's Working Man*, 213–14, 225; Skolnick, *Embattled Paradise*, 214–17; Melvin L. Oliver and Thomas M. Shapiro, *Black Wealth/White Wealth: A New Perspective on Racial Inequality* (New York: Routledge, 1995), 136–47.

32. Diggs, *From the Meadows to the Point*, 79. Julius Wilson, *The Bridge over the Racial Divide* (Berkeley and Los Angeles: University of California Press and Russell Sage Foundation, 1999).

33. Bluestone and Harrison, *The Deindustrialization of America*; Freeman, "How Much Has De-Unionization Contributed to the Rise in Male Earning Inequality?"

34. Interracial projects in Dundalk and Turner Station are more likely to emerge from progressive community associations than from local political clubs. See, for example, Bill Gates, "Dundalk Renaissance Corporation Seeks Community Revitalization," *Dundalk Eagle*, January 9, 2003; Jamie Stiehm, "Dundalk's Spirit," *Baltimore Sun*, December 27, 1992, B2; Marge Neal, "Author, History Buffs Focus on Turner's," *Dundalk Eagle*, February 6, 2003; Lisa Goldberg, "Dundalk Hopes to Pioneer Use of Public Workshops to Spur Redevelopment," *Baltimore Sun*, December 6, 2004, 1A, 8A; and Marge Neal, "Developer Is Interested in Buying Dundalk Village Shopping Center," *Dundalk Eagle*, January 20, 2005, 1, 26. Community-based social justice projects like those being organized in Dundalk and Turner Station are discussed in Margaret Weir and Marshall Ganz, "Reconnecting People and Politics," in *The New Majority: Toward a Popular Progressive Politics*, ed. Stanley B. Greenberg and Theda Skocpol (New Haven: Yale University Press, 1997), 149–71.

35. Hine and Thompson, *A Shining Thread of Hope*, 208–34; Elaine McCrate, and Laura Leete, "Black-White Differences Among Young Women, 1977–1986," *Industrial Relations* 33, no. 2 (1994): 168–83. The increased frequency with which black and white women work cooperatively, both in the workplace and in political efforts, is part of an ongoing discussion within women's studies. See, for example, Louisa A. Tilly and Patricia Gurin, eds., *Women, Politics and Change* (New York: Russell Sage Foundation, 1990); and Iris Berger, Elsa Barkley Brown, and Nancy Hewitt, "Intersections and Collision Courses: Women, Blacks, and Workers Confront Gender, Race, and Class," *Feminist Studies* 18 (Summer 1992): 283–326.

36. Truscello, "Expanding Social Capital Networks of Knowledge," 2004; Howard Winant, "Racial Dualism at Century's End," in *The House that Race Built*, ed. Arnold Rampersad and Wahneema Lubiano, 87–115 (New York: Pantheon Books, 1997).

BIBLIOGRAPHY

Amott, Teresa. *Caught in the Crisis: Women and the U.S. Economy Today*. New York: Monthly Review Press, 1993.

Amott, Teresa L., and Julie A. Matthaei. *Race, Gender, and Work: A Multicultural Economic History of Women in the United States*. Boston: South End Press, 1991.

Anderson, Karen. *Wartime Women: Sex Roles, Family Relations, and the Status of Women During World War II*. Westport, Conn.: Greenwood Press, 1981.

———. "Last Hired, First Fired: Black Women Workers During World War II." *Journal of American History* 69 (June 1982): 82–97.

Appelbaum, Eileen, and Rosemary Batt. *The New American Workplace*. Ithaca: Cornell University Press, 1994.

Argersinger, Jo Ann E. *Toward a New Deal in Baltimore: People and Government in the Great Depression*. Chapel Hill: University of North Carolina Press, 1988.

———. *Making the Amalgamated: Gender, Ethnicity, and Class in the Baltimore Clothing Industry, 1899–1939*. Baltimore: Johns Hopkins University Press, 1999.

Baltimore Metropolitan Planning Council. *Report of the Baltimore Metropolitan Planning Council*. Baltimore: Author, March 1982.

Barnett, Donald F., and Robert W. Crandall. *Up from the Ashes: The Rise of the Steel Minimill in the United States*. Washington, D.C.: Brookings Institute, 1986.

Baron, Ava, ed. *Work Engendered: Toward a New History of American Labor*. Ithaca: Cornell University Press, 1991.

Baxandall, Rosalyn, and Linda Gordon. *America's Working Women: A Documentary History, 1600 to the Present*. New York: W. W. Norton, 1995.

Bederman, Gail. *Manliness and Civilization: A Cultural History of Gender and Race in the United States, 1880–1917*. Chicago: University of Chicago Press, 1995.

Bell, Thomas. *Out of This Furnace*. 1941. Reprint, Pittsburgh: University of Pittsburgh Press, 1976.

Bennett, Harrison, and Barry Bluestone. *The Great U-Turn: Corporate Restructuring and the Polarizing of America*. New York: Basic Books, 1988.

Bensman, David, and Roberta Lynch. *Rusted Dreams: Hard Times in a Steel Community*. New York: McGraw-Hill, 1987.

Benson, Susan Porter. *Counter Cultures: Saleswomen, Managers, and Customers in American Department Stores, 1890–1940*. Urbana: University of Illinois Press, 1986.

Berger, Bennet M. *Working-Class Suburb: A Study of Auto Workers in Suburbia*. Berkeley and Los Angeles: University of California Press, 1968.

Berger, Iris, Elsa Barkley Brown, and Nancy Hewitt. "Intersections and Collision Courses: Women, Blacks, and Workers Confront Gender, Race, and Class." *Feminist Studies* 18 (Summer 1992): 283–326.

Berk, Sarah F. *The Gender Factory: The Apportionment of Work in American Households.* New York: Plenum Press, 1985.

Bernard, Jessie. "The Good-Provider Role: Its Rise and Fall." *American Psychologist* 36, no. 1 (1981): 1–12.

Bernard, Richard M. "A Portrait of Baltimore in 1880: Economic and Occupational Patterns in an Early American City." *Maryland Historical Magazine* 69 (Winter 1974): 341–60.

Bernhardt, Annette, Martina Morris, and Mark Hancock. "Women's Gains or Men's Losses? A Closer Look at the Shrinking Gender Gap in Earnings." *American Journal of Sociology* 101, no. 2 (1995): 302–28.Bingham, Clara, and Laura Leedy Gansler. *Class Action: The Story of Lois Jenson and the Landmark Case that Changed Sexual Harassment Law.* New York: Doubleday, 2002.

Blackburn, McKinley, David Bloom, and Richard Freeman. "The Declining Economic Position of Less Skilled American Men." In *A Future of Lousy Jobs,* ed. Gary Burtless, 31–67. Washington, D.C.: Brookings Institution, 1990.

Blewett, Mary H. *Men, Women, and Work: Class, Gender, and Protest in the New England Shoe Industry, 1780–1910.* Urbana: University of Illinois Press, 1988.

Blight, David. *Race and Reunion: The Civil War in American Memory.* Cambridge: Harvard University Press, 2001.

Bluestone, Barry, and Bennett Harrison. *The Deindustrialization of America: Plant Closings, Community Abandonment, and the Dismantling of Business and Industry.* New York: Basic Books, 1982.

Board of World's Fair Managers, Maryland. *Maryland: Its Resources, Industries and Institutions.* Baltimore: Sun Job Printing Office, 1893.

Bodnar, John. *Steelton: Immigration and Industrialization, 1870–1940.* Pittsburgh: University of Pittsburgh Press, 1977.

———. *Workers' World: Kinship, Community, and Protest in an Industrial Society, 1900–1940.* Baltimore: Johns Hopkins University Press, 1982.

———. "Power and Memory in Oral History: Workers and Managers at Studebaker." *Journal of American History* 75, no. 4 (March 1989): 1201–21.

Bonvillain, Dorothy Guy. "Cultural Pluralism and the Americanization of Immigrants: The Role of Public Schools and Ethnic Communities in Baltimore, 1890–1920." Ph.D. diss., American University, 1999.

Boyett, Joseph H., and H. P. Conn. *Workplace 2000: The Revolution Reshaping America.* New York: Dutton, 1992.

Breines, Wini. *Young, White, and Miserable: Growing Up Female in the Fifties.* Boston: Beacon Press, 1992.

Bridges, Lynetric. "Domestic Workers in Baltimore, 1880–1975." Master's thesis, Morgan University, 2002.

Brody, David. *Steelworkers in America: The Nonunion Era.* New York: Harper and Row, 1960.

Brugger, Robert. *Maryland: A Middle Temperament, 1634–1980.* Baltimore: Johns Hopkins University Press, 1989.

Bruno, Robert. *Steelworker Alley: How Class Works in Youngstown.* Ithaca: Cornell University Press, 1999.

Buba, Tony, and Raymond Henderson. *Struggles in Steel: A Story of African-American Steelworkers*. Braddock, Pa.: Braddock Films, 1996.

Buder, Stanley. *Pullman: An Experiment in Industrial Order and Community Planning, 1880–1930*. New York: Oxford University Press, 1967.

Bullard, Robert D., J. Eugene Gribsby II, and Charles Lee, eds. *Residential Apartheid: An American Legacy*. Berkeley and Los Angeles: University of California Press, 1994.

Bureau of Industrial Statistics of Maryland. *Fifth Annual Report of the Bureau of Industrial Statistics of Maryland, 1893–1901.*

Buss, Terry F., and F. Stevens Redburn. *Reemployment After a Shutdown: The Youngstown Steel Mill Closings, 1977–1985*. Youngstown: Center for Urban Studies, Youngstown State University, 1986.

Butler, Elizabeth Beardsley. *Saleswomen in Mercantile Stores: Baltimore, 1909*. New York: Russell Sage Foundation, 1912.

Byington, Margaret. *Homestead: The Households of a Mill Town*. 1910. Reprint, Pittsburgh: University of Pittsburgh Press, 1974.

Chafe, William. *The American Woman: Her Changing Social, Economic, and Political Roles, 1920–1970*. New York: Oxford University Press, 1972.

Chafe, William H., Raymond Gavins, and Robert Korstad, eds. *Remembering Jim Crow: African Americans Tell About Life in the Segregated South*. New York: New Press, 2001.

Clark-Lewis, Elizabeth. *Living In, Living Out: African American Domestics in Washington, D.C., 1910–1940*. Washington, D.C.: Smithsonian Institution Press, 1994.

Cobble, Dorothy Sue. "Rethinking Troubled Relations Between Women and Unions: Craft Unionism and Female Activism." *Feminist Studies* 16 (Fall 1990): 519–48.

———, ed. *Women and Unions: Forging a Partnership*. Ithaca: Cornell University Press, 1993.

Cohen, Lizabeth. *Making a New Deal: Industrial Workers in Chicago, 1919–1939*. New York: Cambridge University Press, 1990.

———. *A Consumers' Republic: The Politics of Mass Consumption in Postwar America*. New York: Random House, 2003.

Coleman, Richard M. *Wide Awake at 3:00 A.M.: By Choice or By Chance?* New York: W. H. Freeman and Company, 1986.

Collins, Patricia Hill. *Black Feminist Thought: Knowledge, Consciousness, and the Politics of Empowerment*. London: HarperCollins Academic, 1990.

"Communities of Women." Special issue, *Signs: Journal of Women in Culture and Society* 10 (Summer 1985).

Cook, Alice H., Val R. Lorwin, and Arlene Kaplan Daniels. *The Most Difficult Revolution: Women and Trade Unions*. Ithaca: Cornell University Press, 1992.

Coontz, Stephanie. *The Way We Never Were: American Families and the Nostalgia Trap*. New York: Basic Books, 1992.

Cooper, Patricia. *Once a Cigar Maker: Men, Women, and Work Culture in American Cigar Factories, 1900–1919*. Urbana: University of Illinois Press, 1987.

Cowan, Ruth Schwartz. *More Work for Mother: The Ironies of Household Technology from the Open Hearth to the Microwave*. New York: Basic Books, 1983.

Crandall, Robert W. *The U. S. Steel Industry in Recurrent Crisis: Policy Options in a Competitive World*. Washington, D.C.: Brookings Institution, 1981.

Cunningham, James, and Nadja Zalozar. "The Economic Progress of Black Women, 1940–1980: Occupational Distribution and Relative Wages." *Industrial and Labor Relations Review* 45, no. 3 (1992): 540–55.

Davidson, Cathy N., and Bill Bamberger. *Closing: The Life and Death of an American Factory*. New York: W. W. Norton, 1998.

Davis, Mike. *Prisoners of the American Dream: Politics and Economy in the History of the U.S. Working Class*. New York: Verso, 1986.

de Grazia, Victoria. *The Sex of Things: Gender and Consumption in Historical Perspective*. Berkeley and Los Angeles: University of California Press, 1996.

Deaux, Kay, and Joseph C. Ullman. *Women of Steel: Female Blue-Collar Workers in the Basic Steel Industry*. New York: Praeger Publishers, 1983.

Dickerson, Dennis. *Out of the Crucible: Black Steelworkers in Western Pennsylvania, 1875–1980*. Albany: State University of New York Press, 1986.

Diggs, Louis S. *From the Meadows to the Point: The Histories of the African American Community of Turners Station and What Was the African American Community in Sparrows Point*. Baltimore: Uptown Press, 2003.

di Leonardo, Micaela. *The Varieties of Ethnic Experience: Kinship, Class, and Gender among California Italian-Americans*. Ithaca: Cornell University Press, 1984.

———. "The Female World of Cards and Holidays: Women, Families, and the Work of Kinship." *Signs: Journal of Women in Culture and Society* 12, no. 3 (1987): 246–61.

Diner, Hasia R. *Erin's Daughters in America: Irish Immigrant Women in the 19th Century*. Baltimore: Johns Hopkins University Press, 1983.

Directory of Major Manufacturers. Baltimore: Greater Baltimore Committee, 1981.

Douglas, Susan J. *Where the Girls Are: Growing Up Female with the Mass Media*. New York: Three Rivers Press, 1994.

D'Souza, Dinesh. *The End of Racism: Principles for a Multiracial Society*. New York: Free Press, 1995.

Dublin, Thomas. *Women at Work: The Transformation of Work and Community in Lowell, Massachusetts, 1826–1860*. New York: Columbia University Press, 1979.

———. *When the Mines Closed: Stories of Struggles in Hard Times*. Ithaca: Cornell University Press, 1998.

Du Bois, W. E. B. *The Souls of Black Folks*. 1903. Reprint, New York: Barnes and Noble, 2003.

Dundalk-Patapsco Neck Historical Society. "Dundalk, Then and Now, 1894–1980." Booklet. Baltimore: Cavanaugh Press, 1973.

———. "The Neck, 1672–1933." Booklet. Baltimore: Cavanaugh Press, 1973.

———. "Reflections: Sparrows Point, Maryland, 1887–1975." Booklet. Baltimore: Cavanaugh Press, 1976.

Durkin, Barbara Wernecke. *Oh, You Dundalk Girls, Can't You Dance the Polka?* New York: William Morrow, 1984.

Ehrenreich, Barbara, and Diedre English. *For Her Own Good: 150 Years of the Experts' Advice to Women*. Garden City, N.Y.: Anchor Press, 1979.

———. *Nickled and Dimed: On (Not) Getting by in America*. New York: Henry Holt, 2001.

Elfenbein, Jessica I. *The Making of a Modern City: Philanthropy, Civic Culture, and the Baltimore YMCA*. Gainsville: University Press of Florida, 2001.

Erickson, Ethel. "Women's Employment in the Making of Steel, 1943." *Bulletin of the Women's Bureau*, nos. 192–95. Washington, D.C.: Government Printing Office, 1944.

Etter-Lewis, Gwendolyn. *My Soul Is My Own: Oral Narratives of African American Women in the Professions*. New York: Routledge, 1993.

Evans, Sara. *Tidal Wave: How Women Changed America at Century's End*. New York: Free Press, 2003.

Ferree, Myra Marx. "Working-Class Jobs: Paid Work and Housework as Sources of Satisfaction." *Social Problems* 23 (April 1976): 431–41.

Fields, Barbara. *Slavery and Freedom on the Middle Ground: Maryland During the Nineteenth Century*. New Haven: Yale University Press, 1985.

Filene, Peter Gabriel. *Him/Her/Self: Sex Roles in Modern America*. New York: Harcourt Brace, 1975.

Fitch, John A. *The Steel Workers*. 1910. Reprint, Pittsburgh: University of Pittsburgh Press, 1989.

Foltz, Paula R. "Turner Station: A Community Moves Forward." Unpublished manuscript, April 13, 2000.

Fonow, Mary Margaret. "Occupation/Steelworker: Sex/Female." In *Feminist Frontiers*, ed. Laurel Richardson and Verta Taylor, 217–22. New York: McGraw-Hill, 1984.

———. *Union Women: Forging Feminism in the United Steelworkers of America*. Minneapolis: University of Minnesota Press, 2003.

Fox, Karen D., and Sharon Y. Nickols. "The Time Crunch: Wife's Employment and Family Work." *Journal of Family Issues* 4 (March 1983): 61–82.

Franklin, Donna. *Ensuring Inequality: The Structural Transformation of the African American Family*. New York: Routledge, 1997.

Freeman, Richard. "How Much Has De-Unionization Contributed to the Rise in Male Earning Inequality?" In *Uneven Tides: Rising Inequality in America*, ed. Shelton Danziger and Peter Gottschalk, 100–121. New York: Russell Sage Foundation, 1993.

Frisch, Michael. *A Shared Authority: Essays on the Craft and Meaning of Oral and Public History*. Albany: State University of New York Press, 1990.

Frisch, Michael, and Milton Rogovin. *Portraits in Steel*. Ithaca: Cornell University Press, 1993.

Fuchs, Victor R. *How We Live*. Cambridge: Harvard University Press, 1983.

Gabaccia, Donna. *From the Other Side: Women, Gender, and Immigrant Life in the U.S., 1820–1990*. Bloomington: Indiana University Press, 1994.

Gabin, Nancy F. *Feminism in the Labor Movement: Women and the United Auto Workers, 1935–1975*. Ithaca: Cornell University Press, 1990.

Gates, Bill. "Dundalk Renaissance Corporation Seeks Community Revitalization." *Dundalk Eagle*, January 9, 2003.

Geertz, Clifford. "Thick Description: Toward an Interpretive Theory of Culture." In *The Interpretation of Cultures: Selected Essays* (New York: Basic Books, 1973).

George, Rosemary Marangoly, ed. *Burning Down the House: Recycling Domesticity*. Boulder, Colo.: Westview Press, 1998.

Giddings, Paula. *When and Where I Enter: The Impact of Black Women on Race and Sex in America*. New York: William Morrow, 1984.

Glickman, Lawrence B. *A Living Wage: American Workers and the Making of Consumer Society*. Ithaca: Cornell University Press, 1997.

Gluck, Sherna Berger. *Rosie the Riveter Revisited: Women, the War, and Social Change.* New York: Twayne, 1987.

Goldberg, Lisa. "Dundalk Hopes to Pioneer Use of Public Workshops to Spur Redevelopment." *Baltimore Sun*, December 6, 2004, 1A and 8A.

Goldin, Claudia Dale. "Family Strategies and the Family Economy in the Late Nineteenth Century: The Role of Secondary Workers." In *Philadelphia: Work, Space, Family and Group Experience in the Nineteenth Century, Essays Toward an Interdisciplinary History of the City*, ed. Theodore Hershberg, 277–310. New York: Oxford University Press, 1981.

Goodnow, Jacqueline J., and Jennifer M. Boeves. *Men, Women and Household Work.* New York: Oxford University Press, 1994.

Gordon, Ann D., Bettye Collier-Thomas, John H. Bracey, Arlene Voski Avakian, and Joyce Avrech Berkman, eds. *African American Women and the Vote, 1837–1965.* Amherst: University of Massachusetts Press, 1997.

Gordon, Linda. *Heroes of Their Own Lives: The Politics and History of Family Violence.* New York: Viking Penguin, 1988.

Grossman, Hidreth Y., and Nia Lane Chester. "Work, Culture, and Society in Industrializing America, 1815–1919." *American Historical Review* 78 (June 1973): 561.

———. *The Black Family in Slavery and Freedom, 1750–1925.* New York: Pantheon, 1976.

———, eds. *The Experience and Meaning of Work in Women's Lives.* Hillsdale, N.J.: Lawrence Erlbaum Associates, 1990.

Hall, John R., ed. *Reworking Class.* Ithaca: Cornell University Press, 1997.

Halle, David. *America's Working Man.* Chicago: University of Chicago Press, 1984.

Hallock, Margaret. "Unions and the Gender Wage Gap." In *Women and Unions: Forging a Partnership*, ed. Dorothy Sue Cobble, 27–42. Ithaca: Cornell University Press, 1993.

Hanson, Susan, and Geraldine Pratt. *Gender, Work, and Space.* New York: Routledge, 1995.

Hareven, Tamara K. *Family Time and Industrial Time: The Relationship Between the Family and Work in a New England Industrial Community.* New York: Cambridge University Press, 1982.

———. "Review Essay: Origins of the Modern Family in the United States." *Journal of Social History* 17 (1983): 341–56.

Hareven, Tamara K., and Randolph Langenbach. *Amoskeag: Life and Work in an American Factory City.* New York: Pantheon, 1978.

Harris, William. *The Harder We Run: Black Workers Since the Civil War.* New York: Oxford University Press, 1982.

Hartmann, Heidi. "The Family as the Locus of Gender, Class and Political Struggle: The Example of Housework." *Signs: Journal of Women in Culture and Society* 6 (Spring 1981): 366–94.

———. "Roundtable on Affirmative Action and Pay Equity." In *Women and Unions: Forging a Partnership*, ed. Dorothy Sue Cobble, 43–49. Ithaca: Cornell University Press, 1993.

Hartmann, Susan. *The Home Front and Beyond: American Women in the 1940s.* Boston: Twayne, 1982.

———. "Women's Employment and the Domestic Ideal in the Early Cold War

Years." In *Not June Cleaver: Women and Gender in Postwar America, 1945–1960*, ed. Joanne Meyerowitz. Philadelphia: Temple University Press, 1994.

Harvey, David. *Consciousness and the Urban Experience*. Baltimore: Johns Hopkins University Press, 1985.

Hawkins, Richard. "The Baltimore Canning Industry and the Bahamian Pineapple Trade, c.1865–1926." *Maryland Historian* 26 (Fall/Winter 1995): 1–22.

Hayward, Mary Ellen, and Charles Belfoure. *The Baltimore Rowhouse*. New York: Princeton Architectural Press, 2001.

The Heat: Steelworker Lives and Legends. Mena, Ark.: Cedar Hill Publications, 2002.

Hessen, Robert. *Steel Titan: The Life of Charles M. Schwab*. Pittsburgh: University of Pittsburgh Press, 1990.

Hewitt, Nancy A. "Beyond the Search for Sisterhood: American Women's History in the 1980's." *Social History* 10 (October 1985): 306–7.

———. "'The Voice of Virile Labor': Labor Militancy, Community Solidarity, and Gender Identity Among Tampa's Latin Workers, 1880–1921." In *Work Engendered: Toward a New History of American Labor*, ed. Ava Baron, 142–67. Ithaca: Cornell University Press, 1991.

Hill, Michael. "Down and Out in Dundalk." *Baltimore Sun*, December 27, 1992.

Hilson, Robert, Jr. "Osceola Smith Obituary." *Baltimore Sun*, November 23, 1997.

Hine, Darlene Clark. *Hine Sight: Black Women and the Re-Construction of American History*. Bloomington: Indiana University Press, 1994.

Hine, Darlene Clark, and Kathleen Thompson. *A Shining Thread of Hope: The History of Black Women in America*. New York: Broadway Books, 1998.

Hinshaw, John. *Steel and Steelworkers: Race and Class Struggle in Twentieth-Century Pittsburgh*. Albany: State University of New York Press, 2002.

Hinshaw, John, and Paul Le Blanc, eds. *U.S. Labor in the Twentieth Century: Studies in Working-Class Struggles and Insurgency*. Amherst, N.Y.: Humanity Books, 2000.

Hirschfeld, Charles. *Baltimore, 1870–1900: Studies in Social History*. Baltimore: Johns Hopkins University Press, 194l.

Hochschild, Arlie. *The Second Shift: Working Parents and the Revolution at Home*. New York: Viking, 1989.

———. *The Time Bind: When Work Becomes Home and Home Becomes Work*. New York: Henry Holt, 1997.

Hoerr, John. *And the Wolf Finally Came: The Decline of the American Steel Industry*. Pittsburgh: University of Pittsburgh Press, 1988.

Honey, Maureen. *Creating Rosie the Riveter: Class, Gender, and Propaganda During World War II*. Amherst: University of Massachusetts Press, 1984.

Honey, Michael K. *Black Workers Remember: An Oral History of Segregation, Unionism, and the Freedom Struggle*. Berkeley and Los Angeles: University of California Press, 1999.

Hood, Jane. *Becoming a Two-Job Family*. New York: Praeger, 1983.

Horowitz, Roger, and Rick Halpern. "Work, Race, and Identity: Self-Representation in the Narrative of Black Packinghouse Workers." *Oral History Review* 26, no. 1 (Winter/Spring 1999): 23–43.

Hurt, R. Douglas, ed. *African American Life in the Rural South, 1900–1950*. Columbia: University of Missouri Press, 2003.

"Issue on Work Cultures." Special issue. *Feminist Studies* 11 (Fall 1985).

Johnson, Jennifer. *Getting By On the Minimum: The Lives of Working-Class Women.* New York: Routledge, 2002.

Jones, Gareth Stedman. *Languages of Class.* New York: Cambridge University Press, 1983.

Jones, Jacqueline. *Labor of Love, Labor of Sorrow: Black Women, Work, and the Family from Slavery to the Present.* New York: Basic Books, 1985.

———. "One Big Happy Family?" *Women's Review of Books* 6 (February 1989): 4–6.

———. *American Work: Four Centuries of Black and White Labor.* New York: W. W. Norton, 1998.

Juravich, Tom, and Kate Bronfenbrenner. *Ravenswood: The Steelworkers' Victory and the Revival of American Labor.* Ithaca: Cornell University Press, 1999.

Kaltenbach, Chris. "Brian Wilson Was Just Being Himself: The Personality That Made the Dundalk Bashing DJ Famous Is the Same One That Got Him Fired." *Baltimore Sun,* November 26, 1995.

Kammen, Michael. *Mystic Chords of Memory: The Transformation of Tradition in American Culture.* New York: Vintage Press, 1993.

Kantrowitz, Barbara. "Striking a Nerve." *Newsweek,* October 21, 1991, 34–40. Quoted in *America's Working Women,* ed. Rosalyn Baxandall and Linda Gorden. New York: W. W. Norton, 1995.

Katzman, David M. *Seven Days a Week: Women and Domestic Service in Industrializing America.* New York: Oxford University Press, 1978.

Katznelson, Ira. *City Trenches: Urban Politics and the Patterning of Class in the United States.* New York: Pantheon, 1981.

Keating, W. Dennis. *The Suburban Racial Dilemma: Housing and Neighborhoods.* Philadelphia: Temple University Press, 1994.

Kennedy, David M. *Birth Control in America: The Career of Margaret Sanger.* New Haven: Yale University Press, 1970.

Kessler-Harris, Alice. *A Woman's Wage: Historical Meanings and Social Consequences.* Lexington: University Press of Kentucky, 1990.

———. *Out to Work: A History of Wage-Earning Women in the United States.* New York: Oxford University Press, 1991.

Keynes, Milton. "A Woman's Place." In *Space, Place, and Gender,* ed. Doreen Massey, 191–211. Cambridge, U.K.: Blackwell, 1994.

Kimmel, Michael. *Manhood in America.* New York: Free Press, 1996.

Kirp, David L., John P. Dwyer, and Larry A. Rosenthal. *Our Town: Race, Housing, and the Soul of Suburbia.* New Brunswick: Rutgers University Press, 1995.

Klaczynska, Barbara. "Why Women Work." *Labor History* 17 (Winter 1976): 73–87.

———. *The Shadow of the Mills: Working-Class Families in Pittsburgh, 1870–1907.* Pittsburgh: University of Pittsburgh Press, 1989.

Komarovsky, Mirra. *Blue-Collar Marriage.* New York: Random House, 1962.

Komarovsky, Mirra, and Michael S. Kimmel. *The Unemployed Man and His Family: The Effect of Unemployment Upon the Status of the Man in Fifty-nine Families.* 1927. Reprint, New York: Altamira Press, 2004.

Krause, Paul. *The Battle for Homestead, 1880–1892: Politics, Culture and Steel.* Pittsburgh: University of Pittsburgh Press, 1992.

Lamphere, Louise. *From Working Daughters to Working Mothers: Immigrant Women in a New England Industrial Community.* Ithaca: Cornell University Press, 1987.

Landry, Bart. *Black Working Wives: Pioneers of the American Family Revolution.* Berkeley and Los Angeles: University of California Press, 2000.

Lennon, Mary Claire, and Susan Rosenfield. "Relative Fairness and the Division of Housework: The Importance of Options." *American Journal of Sociology* 100, no. 2 (1994): 506–31.

Letwin, Daniel. *The Challenge of Interracial Unionism: Alabama Coal Miners, 1878–1921.* Chapel Hill: University of North Carolina Press, 1998.

Levine, Lawrence W. *Black Culture and Black Consciousness: Afro-American Folk Thought from Slavery to Freedom.* New York: Oxford University Press, 1977.

Levy, Frank. *Dollars and Dreams: The Changing American Income Distribution.* New York: Norton and Russell Sage Foundation, 1988.

Lewin, Tamar. "Women Are Becoming Equal Providers." *New York Times,* May 11, 1995, A27.

———. "Men Assuming Bigger Share At Home, New Survey Shows." *New York Times,* April 14, 1998.

Lipsitz, George. *Rainbow at Midnight: Labor and Culture in the 1940s.* Chicago: University of Illinois Press, 1994.

Litwack, Leon. *Trouble in Mind: Black Southerners in the Age of Jim Crow.* New York: Random House, 1998.

Long, Gail Porter, producer. *Hometown at War.* Baltimore: Maryland Public Television, 1986.

Lubrano, Alfred. *Limbo: Blue-Collar Roots, White-Collar Dreams.* Waltham, Mass.: Reed Business Information, 2003.

Luttrell, Wendy. "'The Teachers, They All Had Their Pets': Concepts of Gender, Knowledge, and Power." *Signs: Journal of Women in Culture and Society* 18 (Spring 1993): 505–46.

Lynd, Staughton. *The Fight Against Shutdowns: Youngstown's Steel Mill Closings.* San Pedro, Calif.: Singlejack Books, 1982.

Makhijani, Pooja. *How Girls Experience Race in America.* Emeryville, Calif.: Seal Press, 2004.

Maryland Public Television. *A Hometown at War.* 1986.

Maryland Steel Company Papers of Frederick W. Wood Papers (FWW Papers). Accession 884. Hagley Museum and Library. Greenville, Delaware.

Massey, Doreen. *Politics and Method: Contrasting Studies In Industrial Geography.* London: Macmillan, 1984.

———, ed. *Space, Place, and Gender.* Cambridge, U.K.: Blackwell, 1994.

Massey, Doreen, and Richard Meegan, eds. *Politics and Method: Contrasting Studies in Industrial Geography.* London: Macmillan, 1984.

Matera, Vincent L. "Consent Decree on Seniority in the Steel Industry." *Monthly Labor Review* 53 (March 1975): 43–46.

Matthaei, Julie. *An Economic History of Women in America.* New York: Schocken Books, 1982.

Matthews, Glenna. *"Just a Housewife": The Rise and Fall of Domesticity in America.* New York: Oxford University Press, 1987.

Mause, Terri Narrell. "Dundalk Timeline." *Dundalk Eagle,* n.d.

———. "Rosie the Riveter To Be Honored During DCC Program Next Week." *Dundalk Eagle,* March 9, 1995, 1, 18.

———. "A Community Built by Steel." *Dundalk Eagle,* March 1, 2000.

May, Martha. "Bread Before Roses: American Workingmen, Labor Unions, and the Family Wage." In *Women, Work, and Protest*, ed. Ruth Milkman, 1–21. New York: Routledge, 1985.

McAdoo, Hariette. *Black Families*. Thousand Oaks, Calif.: Russell Sage Foundation, 1997.

McCall, Leslie. *Complex Inequality: Gender, Class and Race in the New Economy*. New York: Routledge, 2001.

McCrate, Elaine, and Laura Leete. "Black-White Differences Among Young Women, 1977–1986." *Industrial Relations* 33, no. 2 (1994): 168–83.

McDougall, Harold A. *Black Baltimore: A New Theory of Community*. Philadelphia: Temple University Press, 1993.

Melosh, Barbara, ed. *Gender and American History Since 1890*. New York: Routledge, 1993.

Mercier, Laurie. *Anaconda: Labor, Community, and Culture in Montana's Smelter City*. Urbana: University of Illinois Press, 2002.

Meyerowitz, Joanne, ed. *Not June Cleaver: Women and Gender in Postwar America, 1945–1960*. Philadelphia: Temple University Press, 1994.

Mfume, Kweisi. *No Free Ride: From the Mean Streets to the Mainstream*. New York: Ballantine, 1996.

Milkman, Ruth, ed. *Women, Work, and Protest: A Century of Women's Labor History*. Boston: Routledge, 1985.

———. *Gender at Work: The Dynamics of Job Segregation by Sex During World War II*. Urbana: University of Illinois Press, 1987.

Miller, Mark B. *Baltimore Transitions: Views of an American City in Flux*. Rev. ed. Baltimore: Johns Hopkins University Press, 1999.

Mintz, Steven, and Susan Kellogg. *Domestic Revolutions: A Social History of American Family Life*. New York: The Free Press, 1988.

Modell, John, and Tamara K. Hareven. "Urbanization and the Malleable Household: An Examination of Boarding and Lodging in American Families." *Journal of Marriage and the Family* 35 (August 1973): 467–79.

Modell, Judith. *A Town Without Steel: Envisioning Homestead*. Pittsburgh: University of Pittsburgh Press, 1998.

Montgomery, David. *The Fall of the House of Labor: The Workplace, the State and American Labor Activism, 1865–1925*. Cambridge: Cambridge University Press, 1987.

Moore, George L. "The Old 'Company Store' at Sparrows Point." *Baltimore Sun Magazine*, January 4, 1959.

Morgan, Myffannwy, and Hilda H. Golden. "Immigrant Families in an Industrial City: A Study of Households in Holyoke, 1880." *Journal of Family History* 4 (Spring 1979): 59–68.

Murphy, Kevin, and Finis Welch. "Industrial Change and the Rising Importance of Skills." In *Uneven Tides: Rising Inequality in America*, ed. Shelton Danziger and Peter Gottschalk. New York: Russell Sage Foundation, 1993.

Murphy, Peter F. *Studs, Tools, and the Family Jewels: Metaphors Men Live By*. Madison: University of Wisconsin Press, 2001.

Nawrozki, Joe. "Remembering the Boom Times." *Baltimore Sun*, October 11, 1995, 1B, 4B.

Neal, Marge. "Sparrows Point: Well-Planned Homes Revealed Status by Proximity to Steel Plant." *Dundalk Eagle*, July 15, 1999.

———. "Author, History Buffs Focus on Turner's." *Dundalk Eagle*, February 6, 2003.

———. "Developer Is Interested in Buying Dunalk Village Shopping Center." *Dundalk Eagle*, January 20, 2005, 1, 26.

Nelson, Bruce. *Divided We Stand: American Workers and the Struggle for Black Equality*. Princeton: Princeton University Press, 2001.

Neverdon-Morton, Cynthia. "Black Housing Patterns in Baltimore City, 1885–1953." *Maryland Historian* 16 (Spring/Summer 1985): 25–39.

Northrup, Herbert R. *Negro Employment in Basic Industry: A Study of Racial Policies in Six Industries*. Philadelphia: Wharton School/Industrial Research Unit, 1970.

Nye, Francis Ivan, and Lois Wlapis Hoffman, eds. *The Employed Mother in America*. Chicago: Rand McNally, 1963.

Oestreicher, Richard. "Separate Tribes: Working-Class and Women's History." *Reviews in American History* 19 (1991): 228–31.

O'Farrell, Brigid, and Suzanne Moore. "Unions, Hard Hats, and Women Workers." In *Women and Unions: Forging a Partnership*, ed. Dorothy Sue Cobble, 69–84. Ithaca: Cornell University Press, 1993.

Ogden, Annagret S. *The Great American Housewife: From Helpmate to Wage Earner, 1776–1986*. Westport, Conn.: Greenwood Press, 1986.

Okely, Judith, and Helen Callaway. *Anthropology and Autobiography*. New York: Routledge, 1992.

Oliver, Melvin L., and Thomas M. Shapiro. *Black Wealth/White Wealth: A New Perspective on Racial Inequality*. New York: Routledge, 1995.

Olson, Karen. "Old West Baltimore: Segregation, African-American Culture, and the Struggle for Equality." In *The Baltimore Book: New Views of Local History*, ed. Liz Fee, Linda Shopes, and Linda Zeidman. Philadelphia: Temple University Press, 1991.

Olson, Karen, and Linda Shopes. "Crossing Boundaries, Building Bridges: Doing Oral History Among Working-Class Women and Men." In *Women's Words: The Feminist Practice of Oral History*, ed. Sherna Gluck and Daphne Patai. New York: Routledge, 1991.

Olson, Sherry. *Baltimore: The Building of an American City*. Baltimore: Johns Hopkins University Press, 1980.

Oppenheimer, Valerie Kincade. *The Female Labor Force in the United States: Demographic and Economic Factors Governing Its Growth and Changing Composition*. Westport, Conn.: Greenwood Press, 1976.

Orser, W. Edward. *Blockbusting in Baltimore: The Edmondson Village Story*. Lexington: University of Kentucky Press, 1994.

Palmer, David. *Organizing Shipyards: Union Strategy in Three Northeast Ports, 1933–1945*. Ithaca: Cornell University Press, 1998.

Palmer, Phyllis. *Domesticity and Dirt: Housewives and Domestic Servants in the United States, 1920–1945*. Philadelphia: Temple University Press, 1989.

Pappas, Gregory. *The Magic City: Unemployment in a Working-Class Community*. Ithaca: Cornell University Press, 1989.

Parr, Joy. *The Gender of Breadwinners: Women, Men, and Change in Two Industrial Towns, 1880–1950*. Toronto: University of Toronto Press, 1990.

Patterson, Ted. "The Educational Journey of African Americans in Southeastern Baltimore County." Unpublished manuscript, n.d.

Paul, William George. "The Shadow of Equality: The Negro in Baltimore, 1864–1911." Ph.D. diss., University of Wisconsin, 1972.

Peiss, Kathy. *Hope in a Jar: The Making of America's Beauty Culture.* New York: Henry Holt, 1998.

Peterson, David. "Wife Beating: An American Tradition." *Journal of Interdisciplinary History* 23, no. 1 (Summer 1992): 97–118.

Phillips, Marilyn M., producer. *Port Baltimore.* Baltimore: Maryland Public Television, 1997.

Pleck, Elizabeth. "Two Worlds in One: Work and Family." *Journal of Social History* 10, no. 2 (1976): 178–95.

Pleck, Joseph. *The Myth of Masculinity.* Cambridge: MIT Press, 1983.

———. *Working Wives, Working Husbands.* Beverly Hills, Calif.: Sage, 1985.

Portelli, Alessandro. *The Death of Luigi Trastulli: Form and Meaning in Oral History.* Albany: State University of New York Press, 1991.

Potter, David M. *People of Plenty: Economic Abundance and the American Character.* Chicago: University of Chicago Press, 1954.

Potuchek, Jean L. *Who Supports the Family? Gender and Breadwinning in Dual-Earner Marriages.* Stanford: Stanford University Press, 1997.

Preston, Richard. *American Steel: Hot Metal Men and the Resurrection of the Rust Belt.* New York: Prentice Hall, 1991.

Rabinowitz, Howard. *Race Relations in the Urban South, 1865–1890.* New York: Oxford University Press, 1978.

Rachleff, Peter. *Black Labor in the South: Richmond, Virginia, 1865–1890.* Philadelphia: Temple University Press, 1984.

Reardon, Carol. "Pickett's Charge: History and Memory." *Catoctin History* (Spring 2003): 6–16.

Reskin, Barbara F., and Patricia A. Roos. *Job Queues, Gender Queues: Explaining Women's Inroads into Male Occupations.* Philadelphia: Temple University Press, 1990.

Reutter, Mark. *Sparrows Point: Making Steel; the Rise and Ruin of American Industrial Might.* New York: Summit Books, 1988.

Riley, Glenda. *Divorce: An American Tradition.* New York: Oxford University Press, 1991.

Robinson, Jo Ann Ooiman. *Affirmative Action: A Documentary History.* Westport, Conn.: Greenwood Press, 2001.

Rodriguez, Richard. *The Ironies of Education.* New York: Routledge, 1999.

Roediger, David R. *The Wages of Whiteness: Race and the Making of the American Working Class.* New York: Verso, 1991.

Rogers, Daniel T. *The Work Ethic in Industrial America, 1850–1920.* Chicago: University of Illinois Press, 1978.

Rose, James D. "'The Problem Every Supervisor Dreads': Women Workers at the U.S. Steel Duquesne Works During World War II." *Labor History* 36 (Winter 1995): 24–51.

Rosen, Ellen Israel. *Bitter Choices: Blue-Collar Women In and Out of Work.* Chicago: University of Chicago Press, 1987.

Rosenbloom, Joshua L. *Looking for Work, Searching for Workers: American Labor Markets During Industrialization.* New York: Cambridge University Press, 2002.

Rosenzweig, Roy. *Eight Hours for What We Will: Workers and Leisure in an Industrial City, 1870–1920*. New York: Cambridge University Press, 1983.

Rotundo, E. Anthony. *American Manhood: Transformations in Masculinity from the Revolution to the Modern Era*. New York: Basic Books, 1993.

Rouse, Jacqueline Anne. "Lugenia D. Burns Hope: A Black Female Reformer in the South, 1871–1947." Ph.D. diss., Emory University, 1983).

Rubin, Lillian. *Families on the Fault Line: America's Working Class Speaks About the Family, the Economy, Race, and Ethnicity*. New York: HarperCollins, 1994.

Ruggles, Steven. "The Origins of African-American Family Structure." *American Sociological Review* 59 (February 1994): 136–51.

Russakoff, Dale. "Chasing Work Fractures Lives of Steel Families." *Washington Post*, March 28, 1993, A1, A8.

Ryon, Roderick N. "An Ambiguous Legacy: Baltimore Blacks and the CIO, 1936–1941." *Journal of Negro History* 65 (Winter 1980): 21–38.

———. "'Human Creatures' Lives': Baltimore Women and Work in Factories, 1880–1917." *Maryland Historical Magazine* 83 (Winter 1988): 346–64.

Scharf, Lois. *To Work and to Wed: Female Employment, Feminism, and the Great Depression*. Westport, Conn.: Greenwood Press, 1980.

Scheuerman, William. *The Steel Crisis: The Economics and Politics of a Declining Industry*. New York: Praeger, 1986.

Schwartzman, Milt. "Dundalk's Spirit." *Baltimore Sun*, January 5, 1993.

Segrave, Kerry. *The Sexual Harassment of Women in the Workplace, 1600–1993*. Jefferson, N.C.: McFarland and Company, 1994.

Sennet, Richard, and Jonathan Cobb. *The Hidden Injuries of Class*. New York: W. W. Norton, 1972.

Serrin, William. *Homestead: The Glory and Tragedy of an American Steel Town*. New York: Times Books, 1992.

Shackel, Paul A. *Memory in Black and White: Race, Commemoration, and the Post-Bellum Landscape*. Walnut Creek, Calif.: AltaMira, 2003.

Shammas, Carole. *A History of Household Government in America*. Charlottesville: University of Virginia Press, 2002.

Shaw, Stephanie J. *What A Woman Ought To Be and To Do: Black Professional Women During the Jim Crow Era*. Chicago: University of Chicago Press, 1996.

Shelton, Beth Anne. *Women, Men, and Time: Gender Differences in Paid Work, Housework and Leisure*. New York: Greenwood Press, 1992.

Shindel, Len. "They Acted Like Men and Were Treated Like Men." *Baltimore Sun*, February 17, 1992.

Shopes, Linda. "Oral History and the Study of Communities: Problems, Paradoxes, and Possibilities." *Journal of American History* 89, no. 2 (September 2002): 588–98.

Skeggs, Beverley. *Class, Self, Culture*. New York: Routledge, 2004.

Skold, Karen Beck. "The Job He Left Behind: American Women in the Shipyards During World War II." In *Women, War, and Revolution*, ed. Carol R. Berkin and Clara M. Lovett. New York: Holmes and Meier Publishers, 1980.

Skolnick, Arlene S. *Embattled Paradise: The American Family in an Age of Uncertainty*. New York: Basic Books, 1991.

Smith, Valerie. *Not Just Race, Not Just Gender: Black Feminist Readings*. New York: Routledge, 1998.

Sokoloff, Natalie. *Black Women and White Women in the Professions.* New York: Routledge, 1992.

Stacey, Judith. "Can There Be a Feminist Ethnography?" In *Women's Words: The Feminist Practice of Oral History*, ed. Sherna Gluck and Daphne Patai, 11–120. New York: Routledge, 1991.

Staines, Graham L., and Joseph H. Pleck. *The Impact of Work Schedules on the Family.* Ann Arbor: Institute for Social Research, University of Michigan, Survey Research Center, 1983.

Stein, Judith. *Running Steel, Running America.* Chapel Hill: University of North Carolina Press, 1998.

Stiehm, Jamie. "Dundalk's Spirit." *Baltimore Sun*, December 27, 1992, B2.

———. "'Rosies' Share Stories of Their Riveting Work." *Baltimore Sun*, June 13, 2004, B1, B3.

Strasser, Susan. *Never Done: A History of American Housework.* New York: Pantheon, 1982.

Strohmeyer, John. *Crisis in Bethlehem: Big Steel's Struggle to Survive.* Bethesda, Md.: Adler and Adler, 1986.

Sweeney, Vincent D. *The United Steelworkers of America.* Pittsburgh: United Steelworkers of America, 1956.

Tea, Michelle, ed. *Without A Net: The Female Experience of Growing Up Working Class.* Emeryville, Calif.: Seal Press, 2003.

Tentler, Leslie Woodcock. *Wage-Earning Women: Industrial Work and Family Life in the United States, 1900–1930.* New York: Oxford University Press, 1979.

Terborg-Penn, Rosalyn. "African American Women and the Vote: An Overview." In *African American Women and the Vote, 1837–1965*, ed. Ann D. Gordon, Bettye Collier-Thomas, John H. Tracy, Arlene Voski Avakian, and Joyce Avrech Berkman. Amherst: University of Massachusetts Press, 1997.

Terkel, Studs. *American Dreams: Lost and Found.* New York: Ballantine, 1987.

Thomas, Richard W. *Life for Us Is What We Make It: Building Black Community in Detroit, 1914–1945.* Bloomington: Indiana University Press, 1992.

Tiffany, Paul A. *The Decline of American Steel: How Management, Labor and Government Went Wrong.* New York: Oxford University Press, 1988.

Tilly, Louisa A., and Patricia Gurin, eds. *Women, Politics and Change.* New York: Russell Sage Foundation, 1990.

Trotter, Joe William. *The Great Migration in Historical Perspective: New Dimensions of Race, Class, and Gender.* Bloomington: Indiana University Press, 1991.

———. "African American Fraternal Associations in American History: An Introduction." *Social Science History* 28, no. 3 (Fall 2004): 355–66.

Trotter, Joe W., and Earl Lewis, eds. *African Americans in the Industrial Age: A Documentary History, 1915–1945.* Boston: Northeastern University Press, 1996.

Uhlenberg, Peter. "Cohort Variations in Family Life Cycle Experiences of U.S. Females." *Journal of Marriage and the Family* 36 (May 1974): 284–92.

The Union. August 1892.

U.S. Bureau of Labor. *Housing of the Working People in the United States by Employers.* Bulletin 54. September 1904.

———. *Summary of the Wages and Hours of Labor from the Report on Conditions of Employment in the Iron and Steel Industry in the United States.* Document no. 301. Washington, D.C.: Government Printing Office, 1912.

———. *Accidents and Accident Prevention; from the Report on Conditions of Employment in the Iron and Steel Industry in the United States.* Document no. 110, vol. 4. Washington, D.C.: Government Printing Office, 1913.

———. *Working Conditions and the Relations of Employers and Employees; from the Report on Conditions of Employment in the Iron and Steel Industry.* Document no. 110. Washington, D.C.: Government Printing Office, 1913.

U.S. Census Bureau. Twelfth Census. 1900.

———. Thirteenth Census. 1910.

———. Twenty-second Census. 2000.

U.S. Department of Labor, Women's Bureau. *Baltimore Women War Workers in the Postwar Period.* Washington, D.C.: Government Printing Office, 1948.

Vanneman, Reeve, and Lynn Weber Cannon. *The American Perception of Class.* Philadelphia: Temple University Press, 1987.

Ward, Robert. *Red Baker.* Garden City, N.Y.: Doubleday, 1985.

Ware, Susan. *Holding Their Own: American Women in the 1930s.* Boston: Twayne, 1982.

Ware, Vron, and Les Back. *Out of Whiteness: Color, Politics, and Culture.* Chicago: University of Chicago Press, 2002.

Watson, Jerome R. *The Churches of Turners Station: A Legacy of Faith and Family.* Baltimore: Uptown Press, 2002.

———. *Remembering Our Schools, A History of African-American Education in Turner Station and Sparrows Point.* Baltimore: Turners Station Heritage Foundation, 2004.

Weiner, Lynn Y. *From Working Girl to Working Mother: The Female Labor Force in the United States, 1820–1980.* Chapel Hill: University of North Carolina Press, 1985.

Weir, Margaret, and Marshall Ganz. "Reconnecting People and Politics." In *The New Majority: Toward a Popular Progressive Politics,* ed. Stanley B. Greenberg and Theda Skocpol, 149–71. New Haven: Yale University Press, 1997.

Wesley, Charles. *Negro Labor in the United States, 1850–1925.* New York: Russell and Russell, 1927.

Westwood, Sallie. *All Day, Every Day: Factory and Family in the Making of Women's Lives.* Urbana, Ill.: , 1985.

Wheelock, Jane. *Husbands at Home: The Domestic Economy in a Post-Industrial Society.* New York: Routledge, 1990.

Whiteford, William A., director. *Port Baltimore.* Baltimore: Maryland Public Television, 1993.

Wilentz, Sean. *Chants Democratic: New York City and the Rise of the American Working Class, 1788–1950.* New York: Oxford University Press, 1984.

Wilkie, Jane Riblett. "The Decline of Occupational Segregation Between Black and White Women." In *Research in Race and Ethnic Relations: A Research Annual,* ed. Cora Bagley Marrett and Cheryl Leggon, 4. Fairfax, Va.: JIA Publishers, 1982.

Williams, Christine L. *Gender Differences at Work: Women and Men in Nontraditional Occupations.* Berkeley and Los Angeles: University of California Press, 1989.

———. *Still a Man's World.* Berkeley and Los Angeles: University of California Press, 1995.

Williams, Joan. *Unbending Gender: Why Family and Work Conflict and What To Do About It.* New York: Oxford University Press, 2000.

Willie, Charles Vert. *Black and White Families: A Study in Complementarity.* Bayside, N.Y.: General Hall, 1985.

Willis, Dail. "Sollers Point High's First Graduates Return: Despite Racial Barriers, Class of '49 Perseveres." *Baltimore Sun,* April 25, 1999.

Willis, Paul. "Shop Floor Culture, Masculinity and the Wage Form." In *Working-Class Culture: Studies in History and Theory,* ed. John Clarke, Charles Critcher, and Richard Honson, 185–98. London: Hutchinson, in association with the Centre for Contemporary Culture Studies, University of Birmingham, 1979.

Wilson, Julius. *The Bridge over the Racial Divide.* Berkeley and Los Angeles: University of California Press and Russell Sage Foundation, 1999.

Winant, Howard. "Racial Dualism at Century's End." In *The House that Race Built,* ed. Arnold Rampersad and Wahneema Lubiano, 87–115. New York: Pantheon Books, 1997.

Zeidman, Linda. "Sparrows Point, Dundalk, Highlandtown, Old West Baltimore: Home of Gold Dust and the Union Card." In *The Baltimore Book: New Views of Local History,* ed. Liz Fee, Linda Shopes, and Linda Zeidman. Philadelphia: Temple University Press, 1991.

Zieger, Robert H. *American Workers, American Unions, 1920–1985.* Baltimore: Johns Hopkins University Press, 1986.

INDEX